AF708878

WITHDRAWN

Results and Problems in Cell Differentiation

A Series of Topical Volumes in Developmental Biology

12

Editors

W. Hennig, Nijmegen and J. Reinert, Berlin

Differentiation of Protoplasts and of Transformed Plant Cells

Edited by

J. Reinert and H. Binding

With 24 Figures

Springer-Verlag
Berlin Heidelberg New York Tokyo

Professor Dr. J. REINERT
Institut für Pflanzenphysiologie
Königin-Luise-Straße 12–16
D-1000 Berlin 33

Professor Dr. H. BINDING
Botanisches Institut der Universität
Olshausenstraße 40–60
D-2300 Kiel 1

QH
607
.R4
v.12

ISBN 3-540-16539-8 Springer-Verlag Berlin Heidelberg New York Tokyo
ISBN 0-387-16539-8 Springer-Verlag New York Heidelberg Berlin Tokyo

Library of Congress Cataloging in Publication Data.
Differentiation of protoplasts and of transformed plant cells. (Results and problems in cell differentiation; 12) Includes bibliographies and index. 1. Plant protoplasts. 2. Plant cell differentiation. I. Reinert, J. II. Binding, H. (Horst), 1939– . QK725.D53 1986 581.87′612 86-10218

This work is subject to copyright. All rights are reserved, whether the whole or part of the material is concerned, specifically those of translation, reprinting, re-use of illustrations, broadcasting, reproduction by photocopying machine or similar means, and storage in data banks.

Under § 54 of the German Copyright Law, where copies are made for other than private use, a fee is payable to the publisher, the amount of the fee to be determined by agreement with the publisher.

© by Springer-Verlag Berlin · Heidelberg 1986.
Printed in Germany.

The use of registered names, trademarks, etc. in this publication does not imply, even in the absence of a specific statement, that such names are exempt from the relevant protective laws and regulations and therefore free for general use.

Typesetting, printing, and bookbinding: Brühlsche Universitätsdruckerei, Giessen
2131/3130-543210

Contents

III Development of Protoplast Fusion Products

By R. Nehls, G. Krumbiegel-Schroeren, and H. Binding

IV Molecular Biology of Plant Cell Transformation

By N. S. YADAV (With 3 Figures)

Introduction

H. BINDING and J. REINERT

In collaboration with the first authors of this volume

This volume is devoted to the development of cell clones and plants from manipulated cells: isolated protoplasts, cell fusion bodies, and transformed cells.

Isolated protoplasts represent cells which are liberated from their walls and separated from the differentiation pattern of the organism. Investigations on regeneration from protoplasts provide a better understanding of the process and control of developmental pathways.

Whereas protoplast isolation results in alteration of the state of differentiation of a cell, protoplast fusion is a means for the creation of cells with novel genetic constitution. Fascinating features are (1) to hybridize cells which – unlike gametes – did not derive from meiosis products, (2) to bring together foreign plastids and mitochondria and to investigate their parasexual reactions, and (3) to match genetic traits which had been separated for long periods of evolution.

Highly sophisticated techniques have already been elaborated for the transfer of genes by the use of isolated DNA and gene transfer systems. Highly promising results have already been obtained by the use of Ti plasmids of *Agrobacterium,* but direct DNA transformation is also proving to be useful.

Most of the results in these areas are preliminary and/or limited to a few system. It is the aim of this volume to present the main features, but at the same time to draw attention to problems and perspectives of protoplast regeneration and somatic cell genetics in order to stimulate further investigations.

Results and Problems in Cell Differentiation 12
Differentiation of Protoplasts and of Transformed Plant Cells (Edited by J. Reinert and H. Binding)
© Springer-Verlag Berlin Heidelberg 1986

I Isolation and Regeneration of Protoplasts from Higher Plants

S. C. Maheshwari[1], R. Gill[2], N. Maheshwari[1], and P. K. Gharyal[1]

1 Introduction

The aseptic culture of plant cells has emerged in recent years as a powerful technique not only for the study of cell differentiation, but also for plant improvement and agriculture. During the past few years, the potential of plant cell culture has vastly improved due to the fast emerging technology of isolation, cultivation, and fusion, of protoplasts. The term "protoplast" in this respect refers to the spherical plasmolyzed contents of a plant cell enclosed in the plasmalemma and set free of the covering cell wall by a suitable experimental method. The naked cells so obtained constitute an ideal "free cell" developmental system because protoplasts are separate entities capable of reforming cell walls and regenerating whole plants. This fact is of considerable advantage as not only has clonal, large scale, propagation of desired plants become much more efficient, but being discrete, the protoplasts – especially from haploid plants or cell lines – can be handled like microbes and are well suited to mutagenic treatments and somatic cell genetics.

As later chapters will illustrate, protoplasts of two different plants can, for instance, be fused producing hybrids and cybrids which may be unknown in nature; inclusion of subprotoplasts in fusion experiments can help in producing cybrids; genetically novel plants could also be engineered by the transplantation of plastids, mitochondria or chromosomes, or by DNA-mediated transformation.

Nevertheless, the first essential step towards the purposeful application of protoplast technology to genetic modification, e.g. for crop improvement, comprises the isolation of protoplasts and their regeneration into whole plants. The chief objective of this article is to emphasize these basic aspects of in vitro development. It is here that maximum research effort is required in the next few years since, though much progress has been made, the procedures are to a considerable extent still empirical and, ironically, in cereals and legumes – which are our most important crops – the maximum difficulty has been encountered.

While on the current scene the agricultural interest seems to be overriding, protoplasts are rapidly gaining recognition also as an important research tool in a variety of developmental, physiological, and biochemical investigations (see Galun 1981). These include studies on permeability and transport of ions and sol-

[1] Unit for Plant Cell and Molecular Biology, University of Delhi, 110007 Delhi, India.
[2] Present Address: Bio-Organic Division, Bhabha Atomic Research Centre, Bombay, India.

Results and Problems in Cell Differentiation 12
Differentiation of Protoplasts and of Transformed Plant Cells (Edited by J. Reinert and H. Binding)
© Springer-Verlag Berlin Heidelberg 1986

utes (e.g. Akerman et al. 1983; Cornel et al. 1983; Rahat and Reinhold 1983), photosynthesis (e.g. Heber 1982; Chapman and Hatch 1983; Kaiser and Heber 1983), the mechanism of action of plant hormones (e.g. Hooley 1982; Chang et al. 1983; Nebiolo et al. 1983; Norman et al. 1983), of phytochrome (Kim and Song 1981), and maintenance of totipotency, to cite only a few. However, even for these studies isolation of clean and healthy protoplasts is the first step.

There are several general reviews on protoplasts from the first comprehensive one in 1972 by Cocking to those published in recent years by Rao (1982) and Bhojwani and Razdan (1983), but the last specific treatment of the area covered here was by I.K. Vasil and V. Vasil (1980), and from a technical point of view in a book edited by I.K. Vasil (1984).

2 Isolation of Protoplasts

Since the major emphasis in this volume is on differentiation, the techniques for isolation of protoplasts need to be discussed only very briefly. Basically, isolation of protoplasts involves the removal of the cell wall which is done by one of the two methods, mechanical or enzymatic. In either method, the contents are released into an osmoticum which is essential to prevent protoplasts from bursting. The mechanical method relies on sectioning of the plasmolyzed tissue. Cutting such tissues in a suitable manner, e.g. by a razor blade, can result in instantaneous release of protoplasts (e.g. af. Klercker 1892; Binding 1966). However, far more powerful and convenient is the enzymatic method introduced by Cocking (1960) in which cell wall degrading enzymes – such as pectinases and cellulases – can be employed so that protoplasts are set free by the dissolution of the middle lamella and cell wall. Among pectinases, Macerozyme R-10, obtained from the fungus *Rhizopus,* has been the preparation of choice, but in recent years pectolyase, obtained from *Aspergillus japonicus,* has been used (Nagata and Ishii 1979; Hasezawa et al. 1981; Johnson et al. 1982; Ishii and Mogi 1983; V. Vasil et al. 1983). Among cellulases, Onozuka R-10 is very popular, as also Driselase derived from a basidiomycete, *Irpex lacteus* (e.g. Ahuja et al. 1983a, b; Shekhawat and Galston 1983b). The latter enzyme has been found to yield protoplasts in much shorter time – in some cases in only 1–2 h (Arnold and Eriksson 1976; Brar et al. 1980). In many laboratories, enzyme preparations – as available commercially – are used, but sometimes such preparations may contain impurities which may include other toxic materials of both low and high molecular weight, e.g. enzymes like nucleases. Under such circumstances enzyme preparation may be passed through Sephadex or Biogel columns. Such techniques can help in the removal of low or high molecular weight contaminants, and often such purification has led to dramatic improvement of results (Patnaik et al. 1981; Shekhawat and Galston 1983a, b).

Amongst many factors influencing the process of isolation, the correct choice of the plasmolyticum and its concentration are very important since protoplasts are osmotically very active and burst rather easily. Cocking (1960) introduced the use of sucrose and later Ruesink and Thimann (1965) of mannitol and Eriksson and Jonasson (1969) of sorbitol. Since sugar alcohols are metabolically inert and infuse into protoplasts rather slowly, they have come into common use in recent years. Usually, these substances are employed at a concenteration of 0.4–0.6 M. Investigators generally include Ca^{2+} and Mg^{2+} ions (especially the former) for enhancing the stability of the protoplast preparation. Another substance often included is potassium dextran sulfate, first employed by Takebe and co-workers (1971) for in vitro culture of *Nicotiana tabacum* protoplasts and whose beneficial effect has been attributed to adsorption of phenols due to its polyanionic nature. The pH of the incubation medium is adjusted between 5.2 and 6.0. Although in earlier investigations no special buffer had been used, in recent years, substances such as MES (2-N morpholino ethanesulfonic acid) have come into use (Shepard 1980; V. Vasil and I.K. Vasil 1980; Saxena et al. 1982a; Shekhawat and Galston 1983a, b). Generally, enzyme incubations are done in darkness or diffuse light and at a temperature of 25°–30 °C. However, in certain investigations, the tissue has been deliberately exposed to light during incubation. In a special study on *Petunia,* Binding (1974b) showed that the yield improves significantly if light of about 3000 lx was given.

Since protoplasts isolated by enzymes were first regenerated successfully from tobacco mesophyll cells (Takebe et al. 1971), leaves have continued to be the material of choice, the advantage being that a large number of uniform cells can be obtained. Although many investigators, even today, employ leaves from plants grown in the glass house, or growth chamber, or even from the fields, in recent years axenic shoot cultures have come into popular use. The method, first emphasized by Binding (1974a, b), has two advantages: (1) the material is already aseptic, and (2) protoplasts isolated from such cultures have a high regeneration frequency due to conditioning of the donor tissue for rapid growth. Other workers have employed rapidly dividing cell suspensions, recent examples being provided by investigations on *Pennisetum* and *Panicum* (see I.K, Vasil 1982; V. Vasil et al. 1983).

In vitro culture of plant material has made controlled application of environmental factors possible to investigate their significance for the regenerative potency of the isolated protoplasts. Just to mention a few examples, the role of light emerged from experiments with mosses (Binding 1966) and with *Petunia hybrida* (Binding 1974b); the fact that different sucrose concentrations, below isoosmolar strength, are not critical has been shown in *Petunia* (Binding 1974b); increased regeneration frequencies are obtained with protoplasts from moss protonema grown at low calcium (Saxena and Rashid 1981).

Although in most earlier works, plant tissues have been employed for isolation of protoplasts directly, there have been many recent reports of significant increases in yield and stability of protoplasts by pretreatments given to tissues, especially when these are taken from plants grown in soil, i.e. under ordinary conditions. These include plasmolysis, cold treatment, treatment with growth regulators, such as cytokinins, or mere flotation on normal nutrient medium (see Bhojwani and Razdan 1983). Pretreatment can be given also to whole plants since a number of factors, such as age, temperature, light intensity, or photoperiod can affect not only the internal metabolic status of cells, but also cell wall composition affecting both isolation and subsequent stability of protoplasts.

3 Harvesting, Purification, and Culture Methods for Protoplasts

Prior to culture, the protoplasts have to be freed of the hydrolyzing enzyme and debris by centrifugation. In certain situations, like when the sugar alcohols are used, the protoplasts can be gently pelleted. But if sucrose is employed, they float on the top, although here, too, they can be pelleted later by diluting or replacing the sucrose solution with mannitol or ionic solutions. In recent years, techniques are being further refined. For example, substances, such as Percoll or Ficoll are being employed to provide special systems where protoplasts can be banded at the interphase (Gamborg et al. 1983; Nelson et al. 1983). Density gradient centrifugation employing isoosmotic gradients (Harms and Potrykus 1978; Scowcroft and Larkin 1980; Barbier and Dulieu 1983) can yield more homogenous fractions. Even electrophoresis (Halim and Pearce 1980) has been employed for obtaining pure protoplasts.

The methods used for the culture of protoplasts are basically the same as those employed for tissue and cell cultures, but much greater care and dexterity are necessary due to the fragile nature of the protoplasts and the need of transfers to fresh medium. The latter is necessary for adjustment of osmotic pressure and medium composition so as to aid the regeneration of the cell wall and then allow cell divisions to commence. Special procedures may be required when only a few or single protoplasts have to be cultured, such as when fusion products are to be grown. In general, protoplasts are cultured in one of the two following principal ways: (1) liquid culture, and (2) culture in or on solidified media.

The simplest and most common procedure, insofar as liquid culture is concerned, is to dispense a few millilitres of the protoplast suspension in a small Petri dish: typically, a 2 ml suspension can be dispensed in a 5 cm diameter Petri dish so as to obtain a layer only about 1 mm deep and which will allow adequate aeration. In recent years, plastic multidishes with a number of built-in wells are becoming popular and they are especially convenient when several media may have to be tested. Colonies are allowed to develop in these dishes until they are ready for transfer to new media which are usually solid. Another common procedure is that of dispensing the protoplast suspension in the form of small 40–100 μl drops which may be either erect or hanging (if applied to the inside of the cover of a Petri dish). The multiple drop array (MDA) technique devised by Potrykus and co-workers (1979 b) is especially suitable when a large number of media have to be tested and the material available is limited. By restricting the drop size (e.g. to only 40 μl) and by special manipulative skill (i.e. by employing a square grid template), one can place as many as 49 drops in a single Petri dish, in such a way that each drop represents a medium in which the composition has been varied in respect to at least two constituents. Efforts are being made continuously to restrict medium size in such a manner that single protoplasts can be cultured – such procedures are vital for the isolation of special mutants and heterokaryons. A recent example is provided by the work of Koop et al. (1983), where 10 nl droplets have been employed.

As to culture of protoplasts in/on solid medium, the agar plating technique was first employed by Nagata and Takebe (1971) for tobacco. The special advantage of this method is that development of single colonies can be followed and plating efficiencies determined with an accuracy not possible in liquid cultures. The method has also emerged as ideal for picking somatic cell hybrids (see Chapt. II for more details). Generally, the optimal plating density varies between 10^4 and 2×10^4, the minimal density being about 10^3 ml^{-1} for any success at all. Nevertheless, special modifications, such as the feeder-layer (Raveh and Galun 1975) and coculture (Binding and Nehls 1978; Menczel et al. 1978) techniques, have been devised for plating at lower densities. In recent years, agarose – a special form of agar – has been found to be superior to agar (Shepard 1980; Evola et al. 1983; Shillito et al. 1983). Shepard and co-workers (see Shepard 1980) have devised a special quadrant technique of plating which employs agarose and has become especially popular for work on potato protoplasts.

Good results have been obtained with embedding of protoplasts in droplets of agarose media which are flooded by liquid media. Change of media under these conditions is much better tolerated by the cells than in liquid suspension cultures. Furthermore, protoplasts can be clustered in the droplets at extremely high densities enabling fast regeneration (Binding 1964; Binding and Kollmann 1985).

4 Regeneration of Protoplasts

4.1 Cell Wall Formation

Many investigations have been focussed on the early events accompanying the formation of entire cells (called "plastocytes", Binding 1966) from naked protoplasts and especially on those concerned with cell wall regeneration. Cell wall formation has been under especially intensive investigation also for another reason – that is, protoplasts provide a unique opportunity for studying the general mechanism of cell wall synthesis which has mystified plant biologists for long and is still rather poorly understood. In carefully isolated protoplasts there should be no pre-existing mat of cellulosic or other microfibrils to confuse the investigator and they are, thus, potentially excellent material for studies aimed at answering such questions as the role of Golgi bodies and of microtubules in synthesis of cellulose microfibrils and wall deposition. They could also help resolve the mechanism of synthesis of cellulosic microfibrils and a whole subset of related problems. One of these, for example, is whether particulate assemblies (consisting of mobile enzyme complexes) direct the synthesis of cellulose fibrils, and if so, where exactly are they located, i.e. whether outside the plasmalemma or embedded in it? One wonders also how microfibrils grow, whether they lengthen in both directions (as proposed earlier by certain workers) or are synthesized by a unidirectional end-on process of addition of sugar moieties.

A number of early investigations on cell wall formation relied largely on staining by Calcofluor (a technique introduced for protoplast work by Nagata and Takebe 1970) which binds to cellulosic material comprising β 1–4 and β 1–3 linkages – perhaps through hydrogen bonds – and causes it to fluoresce. On occasions other methods have also been used, e.g. polarization microscopy (Abo El-Nil and Hildebrandt 1976; for a more complete review of the earlier literature see I.K. Vasil and V. Vasil 1980). Some of these methods have since been refined. For example, fluorescence induced by Calcofluor can be quantitated and cell wall deposition now measured with a very high degree of sensitivity (Galbraith 1981). Another substance, Tinapol B.O.P.T., manufactured by Ciba-Geigy Co., has also come into use for a similar purpose (Bilkey and Cocking 1982). Although the primary use of such techniques has so far been to provide a convenient method of monitoring cell wall regeneration and to enable optimization of conditions for the culture of protoplasts, in the past few years a large number of additional physical and chemical techniques, such as gas-liquid chromatography (e.g. Takeuchi and Komamine 1978, 1982) and incorporation of radioactive precursors (e.g. Klein et al. 1981), have come into use. Extensive use has also been made of the electron microscope as well as indirect immunofluorescence (the latter for studying the relationship of microtubules to cell wall synthesis). The electron microscopic studies themselves have utilized a large number of more specialized techniques. Apart from thin sectioning and positive staining (e.g. Cocking 1966; Takebe and Otsuki 1973; Fowke et al. 1974; Davey and Mathias 1979), use has been made of negative staining (Burgess and Linstead 1979; Hughes et al. 1976) and standard freeze-etching techniques (Willison and Cocking 1975; Willison and Grout 1978), the last two again in conjunction with preparation of suitable replicas.

Through various techniques, both biochemical and electron microscopic, combined with additional evidence, such as X-ray crystallography, it has become clear that the freshly isolated protoplasts can be completely rid of the cellulose wall and that the fibrillar material that appears on prolonged culture is indeed cellulose even though it may be only weakly crystalline and has dimensions in the range of subelementary fibres (Herth and Meyer 1977; Klein et al. 1981). Studies on incorporation of radioactive precursors show that synthesis can start within a few minutes (10–20) of culture (Klein et al. 1981; see also Hanke and Northcote 1974). However, there have also been several reports that in culture, particularly in liquid medium, a normal cell wall may not always be made (Asamizu and Nishi 1980; Takeuchi and Komamine 1982). The regenerating protoplasts show a lower degree of polymerization of cell wall components than the mesophyll or suspension cultured cells (Blaschek et al. 1982). Further, the pectin-type polysaccharide material is lost to the liquid medium, apparently because the environment is not stable enough for such materials to be incorporated in ways typical of the chemistry of the ordinary cell wall. These observations are of some significance to those attempting to culture protoplasts and may explain the frequent failures to obtain successful regeneration in liquid cultures (e.g. see Takeuchi and Komamine 1982 for *Vinca rosea*), though normal colony development does occur on solid media.

Recently, investigations on microtubular organization have also been made (Lloyd et al. 1980 on tobacco) employing both electron microscopy of thin sections and negatively stained ghosts, as well as indirect immunofluorescence. It does seem that microtubules are in some way involved in cell wall synthesis since they appear in rather large numbers at this time. A role of microtubules in orientation of cellulose microfibrils has been suggested though the exact mechanism remains to be elucidated.

Several investigators have focussed attention on the problem of whether cell wall synthesis and nuclear divisions are related and whether these processes must proceed in a strict sequence (see Hahne et al. 1983 and references cited therein). Although a few authors have expressed themselves in favour of such a notion (Schilde-Rentschler 1977; see also I. K. Vasil and V. Vasil 1980) and a correlative association cannot be denied, the weight of the current evidence seems to be against an obligatory relationship between these processes. Apart from examples relating to investigations on protoplasts where nuclear divisons have been observed without accompanying cell wall synthesis, evidence has come from inhibitor studies conducted by Meyer and Herth (1978) and Galbraith and Shields (1982) who have shown that 2,6-dichlorobenzonitrile can preferentially inhibit cell wall synthesis without inhibiting nuclear divisions, indicating thereby that the two processes can be separated in space and time.

Finally, it is of interest to enquire as to whether growth regulators stimulate the process of wall formation. It is almost certain that so far as divisions are concerned both auxins and cytokinins are stimulatory and almost invariably required. At an earlier stage, auxins play the more critical role, but after some hours cytokinins become more important (Meyer and Cooke 1979). However, for cell wall synthesis per se, their role is questionable. Even for auxins, despite longstanding ideas about their role in cell wall synthesis or cell expansion (and the frequent general statements that they are necessary for cell wall formation and cell divisions), there does not seem to be sufficient compelling evidence to suggest that they specifically stimulate cell wall formation. Nevertheless, hormones are required at later stages of growth of the protoplast – after the cell wall has been regenerated – and thus they are normal constituents of the culture media.

4.2 Cell Division, Colony Formation, and Differentiation

Mitotic activity in moss plastocytes cultured in pure ionic media (diluted seawater) usually sets in only after reduction of the osmolarity of the media (Binding 1966; Schieder and Wenzel 1972). Most commonly, however, cell divisions occur already in the high osmolar protoplast culture media. A typical series of events starting with mesophyll protoplasts from cell wall regeneration to plant formation is shown in Fig. 1 a–f. Wall formation is followed by an increase in the size of the cell and rearrangement of the chloroplasts which become scant and yellow. The cell then divides equally or unequally. Nagata and Takebe (1971) reported that in tobacco the first mitotic division required 2–5 days. This is true not only in this plant as confirmed by other workers, but also in a majority of others. In some species, bicellular regenerants appeared already within 32 h (e.g. in *Petunia:* Binding 1974b). But in certain plants, e.g. in mesophyll protoplasts of *Arachis hypogaea,* this may take a week or more (Oelck et al. 1982), and in cereals such as rye about 2 weeks (Wenzel 1973). Usually, the required time was shorter when the protoplasts were derived from juvenile cells, e.g. from cell suspensions or from shoot tips.

The various factors controlling growth and differentiation are discussed more fully in the following section. However, a few general remarks may be made here.

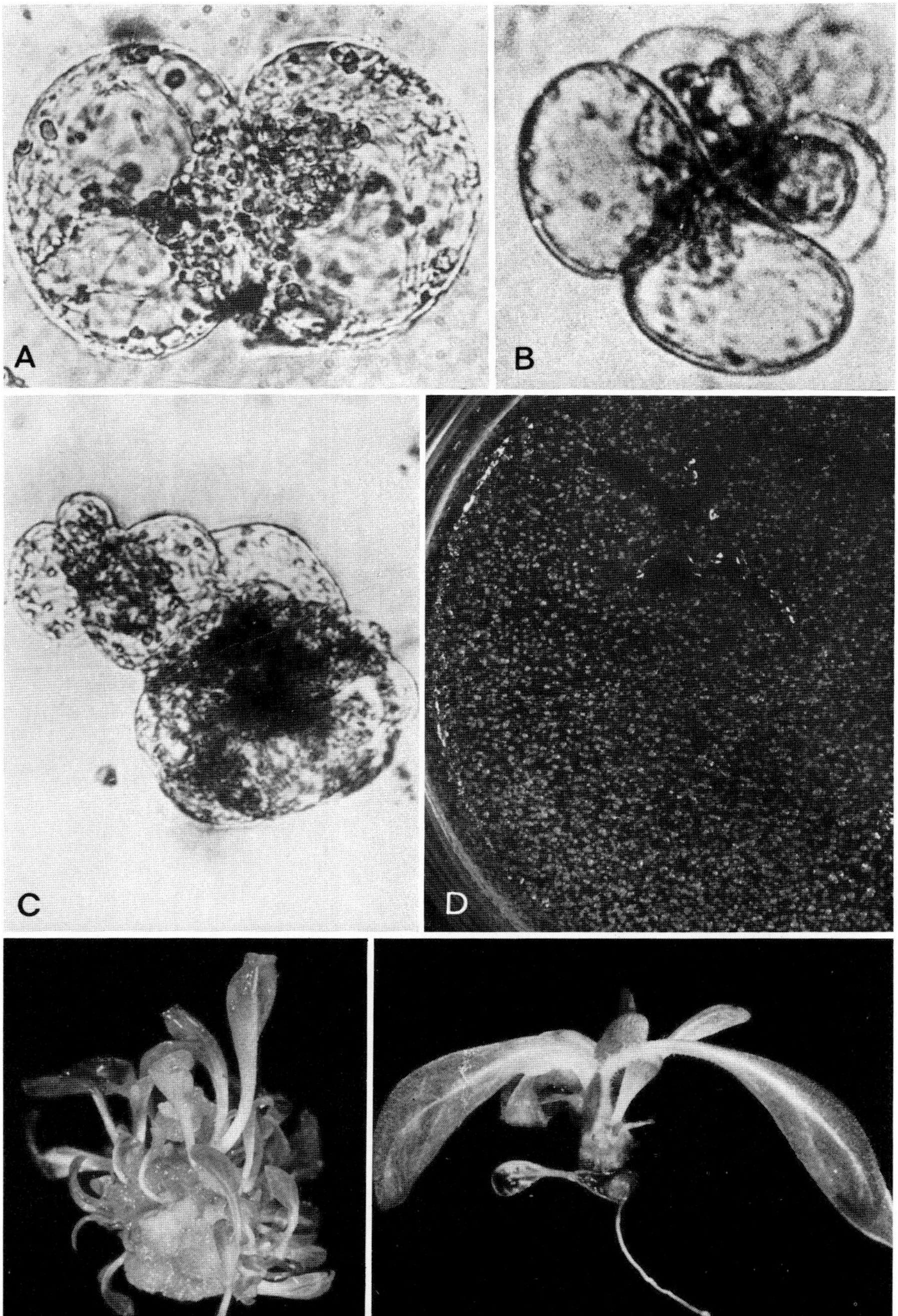

Whereas in some species even organogenesis occurs in high osmolar media (Binding et al. 1981), in a majority of cases the osmotic pressure needs to be gradually reduced by diluting the osmoticum to promote proliferation. In liquid and softagar media, this is relatively easily accomplished by merely adding some more nutrient medium which may contain lowered levels of the osmoticum or none at all. But, when transfer to new medium is desired which is necessary for inducing differentiation (see below), young colonies can be collected by centrifugation or filtration. In cultures in/on solid media transfers to new media are obligatorily required and accomplished by either scraping of colonies or cutting blocks of agar and overlaying them on fresh medium, or transferring agarose containing culture droplets.

The second point concerns growth hormones. It has been stated earlier that auxins and cytokinins are critically required. The combination ensures initiation of repeated mitotic divisions leading to the development of multicellular colonies which in about 3–4 weeks reach a size of 0.1–1.0 mm in diameter and are visible to the naked eye. However, in transfers subsequent to the initial culture – for shoot formation – it is necessary to reduce the level of auxins and even replace the auxin earlier used with one of lower potency. The cytokinin level, however, needs to be continued or even increased. Finally, for rooting, the cultures have again to be transferred to a medium which must usually contain some auxin, but no cytokinin.

After divisions have been initiated, during subsequent transfers, the density too needs to be gradually decreased to avoid overcrowding and the light condition readjusted from that of low intensity to higher intensity usual in plant growth chambers or greenhouses.

It would be apparent that even in the simplest procedures for regeneration of plants from protoplasts at least three media with changes in levels of auxin, cytokinin, and osmoticum – and frequently also in the basal medium itself – must be employed as summarized below (see also Nagata and Takebe 1971).

(1) A regeneration medium suitable for cell wall formation and initiation of cell divisions leading up to the formation of visible colonies – this medium should contain hormones, sugar(s), salts, and vitamins, but especially a high amount of osmoticum, such as mannitol.
(2) A shoot differentiation medium containing a lower concentration of the osmoticum (or none at all) and other addenda – the level of auxin, however, has to be lower and of the cytokinin often higher. If cultures are started in liquid or soft agar medium, colonies should have been by this time, transferred to more solid agar-gelled medium.

Fig. 1 A–F. Different stages in the culture and regeneration of plants from mesophyll protoplasts of *Nicotiana plumbaginifolia* Viv. **A** 6-day-old dividing protoplast on Ohyama and Nitsch's medium supplemented with 2,4-D (1.0 mg l^{-1}), benzylaminopurine (1.0 mg l^{-1}), 14% sucrose, and 0.6% agar. **B** Four-celled stage after 10 days, on the same medium. **C** 4-week-old colony, developed from protoplasts on the same medium. **D** General view of culture showing 1-month-old colonies. **E** 14-week-old callus showing formation of shoots on differentiation medium (MS + IAA 4.0 mg l^{-1} + kinetin 2.56 mg l^{-1}). **F** 22-week-old plantlet differentiated from protoplast-derived callus on MS medium supplemented with IAA (1.0 mg l^{-1}) and kinetin (0.04 mg l^{-1})

(3) Finally, a root differentiation medium containing some auxin, but no cytokinin – the osmoticum, if present, is now omitted altogether.

Another new dimension to studies on culture of protoplasts has been added by the possibility existing now of a deliberate choice of the mode of development of plantlets, i.e. whether through the ordinary course of differentiation via a callus stage, or through embryogenesis, if necessary directly. In several plants somatic embryogenesis is now more or less well established, e.g. *Nicotiana tabacum* (Lörz et al. 1977; Harms et al. 1979), *Atropa belladonna* (Gosch et al. 1975 b), *Antirrhinum* (Poirier-Hamon et al. 1974), carrot (Kameya and Uchimiya 1972; Dudits et al. 1976), various species of *Brassica* (Thomas et al. 1976; Kohlenbach et al. 1982), *Citrus* (Vardi et al. 1975, 1982), legumes, such as *Medicago* (Arcioni et al. 1982; Lu et al. 1983 b) and grasses, such as *Pennisetum americanum* (V. Vasil and I. K. Vasil 1980), *P. purpureum* (V. Vasil et al. 1983), and *Panicum maximum* (Lu et al. 1981). While for certain studies it may not make a serious difference as to which mode is adopted, for others (where it is necessary to be sure of genetic uniformity or origin of plants from a single cell) it may be advantageous to shorten or by-pass the callus phase and have embryoids develop directly from protoplasts (see Krikorian 1982). The information on this aspect is now beginning to accumulate and, hopefully, will allow in the near future precise control of development, omitting one or the other phase completely. Recently, reduction of the unorganized callus phase up to adventitious shoot formation has been obtained by the use of the high density plating technique in agarose media (Binding 1984; Binding and Kollmann 1985).

5 Factors Affecting Regeneration of Protoplasts

5.1 The Chemical Environment

5.1.1 Nutrient Medium and Its Mineral Components

In principle, the nutritional requirements of protoplasts should basically be similar to those of isolated cells. However, many workers have emphasized the need of special care, such as in adjusting medium strength at the early stages of growth. Shepard and Totten (1975, 1977) in their work on potato have especially advocated this need and recommended starting cultures from a low strength nutrient medium only. A similar example is provided by Mühlbach's work on tomato (1980). However, the rationale for these changes is not entirely clear and, sometimes, the methodology adopted even in the same laboratory has not been very consistent (e.g. see Shepard 1980; Shahin and Shepard 1980). From the viewpoint of composition, at least at the usually employed densities between 10^4 to 10^5 protoplasts ml^{-1}, the more common media that have been formulated for callus and cell cultures are satisfactory enough also for protoplasts. Thus, the Nagata and Takebe (1971) medium has been derived from the classical formulation of Murashige and Skoog (1962). Further modifications of these media have led to the development of To (Bourgin et al. 1979), F5 (Frearson et al. 1973), DPD

(Durand et al. 1973), and V-47 (Binding 1974 b) media. Gamborg and co-workers (1968) have extensively employed the B5 nutrient medium. Special mention may be made of the so-called KM medium developed by Kao and Michayluk (1975) originally for *Vicia hajastana*. The medium is unique for its extraordinary complexity, but this was especially designed for cultures at rather low densities where protoplasts ordinarily are unable even to survive. Binding and Nehls (1977) have formulated the V-KM medium which combines the inorganic salts of V-47 and modified organic supplements of KM medium and which according to them is one of the best (Binding et al. 1980, 1981).

The principal modification of the standard culture media with respect to inorganic constituents insofar as specific nutrients are concerned relate to alterations in the levels of ammonium and calcium which have been found to be especially important for regeneration of protoplasts. However, while some investigators have found ammonium ions to have a favourable effect (Arnold and Eriksson 1977; Bhojwani et al. 1977; Nehls 1978; Crepy et al. 1982), including a special one on somatic embryogenesis, e.g. in tobacco (Lörz et al. 1977), others have reported that they are deleterious (Kao et al. 1973; Meyer 1974; Upadhya 1975; Shepard and Totten 1977; Boyes et al. 1980; Boyes and Sink 1981; Zapata et al. 1981), so that a clear consensus is yet to emerge. There are examples where for the same plant, e.g. potato, one group recommends deletion of ammonium (Shepard and Totten 1977), but others include it in the medium (Binding et al. 1978; Thomas 1981; Thomas et al. 1982). In contrast, the beneficial role of increased calcium for the survival and subsequent regeneration of protoplasts is well established in many cases (Kao et al. 1973; Pelcher et al. 1974; Arnold and Eriksson 1977; Bhojwani et al. 1977; Bourgin and Missonier 1978; Caboche 1980; Xuan and Menczel 1980; Jia 1982). Generally, the calcium effect is specific, but in pea mesophyll protoplasts magnesium ions have been reported to partially substitute for calcium (Arnold and Eriksson 1977) Caboche (1980) found that an increase of both calcium and magnesium in To medium was beneficial for culture of tobacco protoplasts at low densities.

Very few studies exist on the role and optimal levels of specific micronutrients in protoplast growth and division – apparently it has been assumed that their levels are adequate and no special changes are called for. In one study on tobacco protoplasts even total omission of minor elements did not make much difference for at least the early phase of growth (Meyer and Abel 1975 b); only iron was critically required. In a later study by Arnold and Eriksson (1977), the importance of iron was further supported and among the different sources of iron – Fe-EDTA, $FeCl_3$ and Fe-citrate – Fe-EDTA was found to be the best. The latter study also focussed on the requirement of zinc which is also critical for the growth and division of protoplasts of pea. However, unlike carrot (Wallin and Eriksson 1973) studied earlier where wall formation in protoplasts occurred in a satisfactory manner only when zinc, like iron, was best supplied in the chelated form, in pea zinc does not have to be supplied in the chelated form. But the studies of Arnold and Eriksson are of greater interest on another count, i.e. for revealing a specific requirement for iodine (supplied as potassium iodide; see, however, Caboche 1980 on tobacco). Though certain media, such as the B5 medium, are normally provided with this element, few investigators have made any specific studies.

It would appear that for most plants the general micronutrient levels in common tissue and cell culture media are satisfactory. However, Caboche's (1980) study of tobacco protoplasts in To medium showed that significant stimulation in plating efficiency occurred if their level was increased about threefold.

5.1.2 Carbon Source

Sucrose (1–4%) has been generally adopted as the carbon source and if protoplasts have been isolated in solutions of mannitol or sorbitol they must eventually be brought into culture media containing some sucrose. In the detailed study on *Nicotiana tabacum,* Uchimiya and Murashige (1976) reported that sucrose supported cell wall regeneration and initiated cell divisions most effectively. Other sugars ranked as follows: cellobiose > glucose > galactose, in order of their effectiveness; whereas cellobiose was nearly as effective as sucrose, galactose was almost useless. Similarly, in *Pisum sativum,* sucrose was found to be essential for the synthesis of cell wall (Arnold and Eriksson 1977). The addition of xylose, arabinose, and glucose had a favourable effect, but apparently could not substitute for sucrose.

However, an important point to take note of is that relative to mannitol or sorbitol, the concentration of sucrose has to be much lower since at a level high enough to serve also as an osmoticum it inhibits cell divisions. This is well illustrated by the studies of Wallin and Eriksson (1973) on carrot who employed various combinations of sucrose and sorbitol and found that a high level of sucrose alone (i.e. to serve both as an osmoticum and a carbon source) was harmful and the best results were obtained when sucrose was combined with sorbitol. An identical conclusion that sucrose alone is harmful, has been drawn by Bhatt and Fassuliotis (1981) in their work on eggplant. The inhibitory effect of sucrose has been emphasized also by Gamborg, Kao, and co-workers (Grambow et al. 1972; Gamborg et al. 1975; Kao and Michayluk 1975, 1980; Brar et al. 1980). In fact, many workers have found that glucose can serve both as an osmoticum and a carbon source, often with much better results, and it has become a general practice to substitute it for mannitol.

The beneficial effect of sugars other than sucrose or glucose, such as ribose and xylose (occasionally their corresponding alcohols) can be ascribed to their role in synthesis of their cell wall components, such as pectins and hemicelluloses and several workers have included them in nutrient media (Constabel et al. 1973; Kao and Michayluk 1975; Simmonds et al. 1979; V. Vasil and I.K. Vasil 1979; Shahin and Shepard 1980). In fact, the KM medium devised by Kao and Michayluk (1975) for culturing protoplasts at low densities contains the following: fructose, ribose, xylose, mannose, rhamnose, and cellobiose, in addition to sorbitol and mannitol. Even though no detailed studies have been published in regard to the roles of individual sugars, their omission as a group clearly results in lowered plating efficiencies.

Apart from sugars, organic acids, such as citric, malic, and fumaric, intermediates of the Krebs cycle, also greatly enhance plating efficiencies (Kao and Michayluk 1975) and are components of the KM medium. A beneficial role of such acids has been recently supported also by studies of Caboche (1980), Negrutiu and Mousseau (1980), and Muller et al. (1983) on *Nicotiana* species, and Ahuja et al. (1983 b) on *Trifolium repens*.

5.1.3 Vitamins

Nagata and Takebe (1970) found that while protoplasts of *Nicotiana tabacum* may commence divisions even in the absence of thiamine and meso-inositol, they do so well only in their presence. A detailed study of vitamin requirement in pea indicated the need for nicotinic acid, pyridoxine, and thiamine (Arnold and Eriksson 1977). Inclusion of folic acid and meso-inositol also gave a high division frequency, but biotin, Ca-pantothenate, and riboflavin had no effect. In oat, the addition of biotin was essential for initiation of divisions (Brenneman and Galston 1975), although no laboratory has been able to show sustained colony growth so far. The critical requirement of meso-inositol was recently confirmed by Caboche (1980) in tobacco and by Xu et al. (1981) for *Phaseolus aureus*. However, excessive concentrations of vitamins are to be avoided, as for protoplasts of *Antirrhinum majus,* an increased concentration of vitamins resulted in the formation of giant cells and abnormal proembryos (Poirier-Hamon et al. 1974).

5.1.4 Growth Regulators and Other Supplements

Whereas external phytohormones may not be required in bryophytes (Binding 1966; Schieder and Wenzel 1972), they are apparently essential in the higher embryophytes. In their classical work on the culture of tobacco protoplasts, Nagata and Takebe (1970) showed that auxin and cytokinins were essential for initiating cell divisions and further development of protoplasts. This has been thoroughly documented not only in tobacco (e.g. Uchimiya and Murashige 1976), but also in many other plants. An auxin, in particular, is required from the beginning of the culture (a *Citrus* line is, however, an interesting exception. Vardi et al. 1975, 1982; see also Kohlenbach et al. 1982 for *Brassica*). Though the role of auxin in cell wall formation has not been substantiated adequately, its role in initiating cell division in the vast majority of plants is beyond doubt. In fact, in some systems, such as tobacco (Uchimiya and Murashige 1976), carrot (Grambow et al. 1972), soybean (Kao et al. 1971), and corn (Potrykus et al. 1977) the presence of an auxin alone can induce mitotic activity. Commonly a range between 0.2 and 2.0 mg l^{-1} or 10^{-6}–10^{-5} M is employed. However, detailed studies suggest that such concentrations are inhibitory after only a few days of initial culture owing to a toxic effect, especially if protoplasts are cultured at a low density (Caboche 1960; see also Engler and Grogan 1983). Continuing high levels are also inhibitory for shoot organogenesis and by timely transfer to low auxin medium, the total period required for plantlet formation can be reduced in tobacco from a normal of 12–14 weeks to only 7–8 weeks (Wernicke and Thomas 1980).

Although it is customary to add cytokinins along with auxins, it has been demonstrated that for tobacco mesophyll protoplasts the cytokinin is required later, 20–24 h before mitosis begins (Meyer and Cooke 1979). In any event, other plants like *Antirrhinum,* pea and carrot seem to need cytokinins obligatorily and from the beginning to induce divisions (Poirier-Hamon et al. 1974; Dudits et al. 1976;

Arnold and Eriksson 1977). However, the optimum concentration of both auxins and cytokinins may vary somewhat depending upon a number of factors, such as the density of the inoculum, source of protoplasts, and genetic factors, since considerable variation is encountered even within the same genus (Izhar and Power 1977).

By and large, the usefulness of growth regulators has been restricted so far to the auxins and cytokinins, already well-known for their growth promoting effects. However, following certain reports on cultured tissues of potato, Shepard (1980) found that abscisic acid enhances shoot formation in protoplast-derived cells and Picloram is especially effective for low density cultures (Muller et al. 1983). Notable also are the reported effects of another class of compounds, the polyamines. Galston and co-workers (1980; see also Kaur-Sawhney et al. 1980) first reportedly found stimulatory effect of such compounds on DNA synthesis and mitosis in oat protoplasts. Recently, a detailed study by Huhtinen et al. (1982) on *Alnus glutinosa* and *A. incana* has shown that ornithine and putrescine are indeed very effective in supporting cell divisions and colony formation.

Various other growth promotive supplements have also been employed for the culture of protoplasts, namely, amino acids and their amides, casein hydrolysate, coconut milk, extracts of yeast, malt, and lately also of potato. Of the amino acids and their derivatives, glutamine has been found to be markedly beneficial for protoplasts of several plants, such as *Datura innoxia* (Furner et al. 1978), *Pseudotsuga menziessi* (Kirby and Cheng 1979), *Cichorium intybus* (Crepy et al. 1982), *Glycine soja* and *G. tabacina* (Gamborg et al. 1983), *Solanum viarum* (Kowalczyk et al. 1983), glutamine and asparagine for *Trigonella foenum-graecum* (Shekhawat and Galston 1983a), glutamine and serine for *Salpiglossis sinuata* (Boyes and Sink 1981), all three, glutamine, asparagine, and serine for *Vicia narbonensis* (Donn 1978), and glutamine, arginine, glycine, and aspartic acid for *Rehmannia glutinosa* (Xu and Davey 1983).

For various reasons chemically defined media are preferable over extracts. Nevertheless, in difficult situations, use of extracts and conditioned media has permitted some progress to be made – an early example is that of soybean (Kao et al. 1970). Thus, the addition of casein hydrolysate, yeast extract, and coconut milk, has been found to enhance divisions of protoplasts derived from mesophyll tissue as well as from cultured cells of a number of plants. In a recent investigation on *Brassica napus,* Kohlenbach et al. (1982) found potato extract to be extremely useful for regenerating plants from protoplasts. In this plant, despite considerable progress in use of tissue cultures on other fronts – such as successful induction of haploidy and regeneration of callus from protoplasts – differentiation of plants is not easy. However, success in this regard has been obtained employing protoplasts of stem embryos and a protocol which again requires potato extract.

Another substance which has recently been found to enhance plating efficiency is activated charcoal (Kohlenbach et al. 1982; Carlberg et al. 1983). Charcoal has the property of adsorbing inhibitors and increasing use may be found of this substance for research on protoplast culture.

5.2 The Physical Environment

5.2.1 Density

For successful culture of protoplasts, density is a very critical factor. Many studies show that the optimal density generally is between 10^4 and 10^5 protoplasts ml^{-1}, the exact value depending on the plant species and such factors as the tissue employed and the physiological condition of the donor plant. Very high densities are detrimental obviously on account of overcrowding and competition for nutrients.

However, the reasons for inability of protoplasts to grow at low densities and the existence of a cooperative effect are not understood clearly. The general belief is that at low densities protoplasts lose considerable amounts of vital substances to the medium – a situation which may not be entirely compensated by uptake of fresh nutrients. However, other effects cannot be ruled out entirely. A suggestion has been made, for example, for a "detoxification" mechanism as well – a larger population of protoplasts may be able to detoxify a deleterious substance in the immediate environment (Kao and Michayluk 1975; Shepard and Totten 1975).

In recent years, some success has been obtained in culturing protoplasts at low densities by the feeder-layer, nurse culture, and other techniques (Raveh et al. 1973; Binding and Nehls 1978; Gleba 1978). Kao and Michayluk (1975) developed a complex medium supplemented with several sugars, organic acids, amino acids, nucleic acid bases, vitamins, and various growth regulators, such as zeatin and NAA in addition to 2,4-D. With the use of such medium, protoplasts have been grown at a density of 25–50 ml^{-1}. The density could be further lowered to only 1–2 ml^{-1} if casamino acids and coconut milk were added to the medium.

Caboche (1980) recently carried out a rather detailed investigation on haploid protoplasts of tobacco and found that although protoplasts could not be raised on a simple defined medium when cultured at low density from the very beginning, they could in fact be grown successfully in such a medium at a density as low as 1–4 ml^{-1}, provided that they had been precultured at a normal, higher, density for just 4 days and subsequently a reduction was made in the level of the auxin, together with some modifications in the To medium (such as addition of glutamine).

5.2.2 pH

A pH in the range of 5.5–5.8 is satisfactory for the culture of most protoplasts. However, pH values somewhat above 6.0 also markedly enhance cell divisions in protoplasts of pea (Gamborg et al. 1975), cowpea (Bharal and Rashid 1980), and *Asparagus officinalis* (Mackenzie et al. 1973).

Generally, the pH of the medium is adjusted before autoclaving (unless use is made of Seitz or other filtering devices). But it is a common experience that autoclaving changes the pH of the medium significantly. Further, growing protoplasts themselves change the pH of the medium. Recently, Roscoe and Bell (1981) have recommended the use of a pH indicator, bromocresol purple, which can be added to the culture medium to monitor changes in pH. This dye is autoclavable and apparently does not affect the development of protoplasts of *Petunia*.

5.2.3 Temperature

The temperature employed for culture of protoplasts has generally ranged between 22° and 28 °C. However, at either extremes, sensitivity may be high. Thus, Zapata et al. (1977) found that whereas in both *Lycopersicon esculentum* and *L. peruvianum* divisions proceeded almost equally well between 27° and 29 °C, at 25 °C the protoplasts of *L. peruvianum* failed to divide at all and at 31 °C such was the case for *L. esculentum* (2% plating efficiency), though the former species was not much affected.

A far more striking effect of temperature, again in tomato, has been demonstrated recently by Mühlbach and Thiele (1981). Chilling of freshly isolated mesophyll protoplasts at 7 °C for 12 h in the dark enhances the division frequency by more than twofold. The effect has been ascribed to the excretion of some factor by the protoplasts into the medium which stimulates cell division, since replacement of the medium with fresh medium nullified the promotive effect of chilling. Recently, it has been found that almost similar effects as described above can be had by chilling of the donor tissue. Thus, a considerable increase has been obtained in division frequency of protoplasts of the legume, *Cyamopsis tetragonoloba*, if the source tissue, i.e. the cotyledons, were kept at 10 °C before isolation of protoplats (Saxena et al. 1982a). Similar pretreatment of tissue to low temperature has also been employed by Engler and Grogan (1983), Ahuja et al (1983b) and Bidney et al. (1983). One is reminded also of the strong promotive effects of chilling on production of haploid plants by pollen cultures (see Maheshwari et al. 1982), but almost nothing is known of the biochemistry of such effects.

5.2.4 Light

Attention on appropriate light conditions during cell wall regeneration was already emphasized in the early investigations on moss protoplasts (Binding 1966). Optimum light intensity for *Funaria* was about 1000 lx, lower intensities resulted in reduced osmotic stability and higher intensities lead to extensive budding. Budding occurrred also when a day/night regime was applied. Nagata and Takebe (1971) were the first to comment on the role of light on the division of protoplasts of higher plants. For protoplasts from leaves of *Nicotiana tabacum* cv. Xanthi, the best results were obtained at a relatively high intensity of 2300 lx, rather than at 700 or at 5000 lx, when the colonies were less green. Somewhat contrary to their result, in another more detailed investigation on *Nicotiana tabacum* cv. Samsun, where light conditions were varied even as the protoplasts underwent divisions, a considerable increase in plating efficiency was obtained if cultures were initially kept at a low intensity of 400 lx for 48 h and later transferred to 3000 lx (Enzmann-Becker 1973). A perusal of several later reports indicate that in many plants high intensity light is indeed inhibitory if given from the very beginning, such an effect arising probably from the bleaching of chloroplasts. Presumably, it is for this reason that in many laboratories it has been a common practice to keep the cultures initially, i.e. for the first 12–48 h (Chupeau et al. 1974; Potrykus et al. 1979a; Thomas et al. 1976 for *Brassica napus*) or a few days (Ahuja et al. 1983a for *Lotus corniculatus;* Muller et al. 1983 for various species of *Nicotiana* and *Petunia*; Schenk and Hoffmann 1979 for *Brassica napus*) or even a whole month (Mühlbach 1980 for tomato), in darkness or dim light and only subsequently return them to stronger light of about 2000–3000 lx or more.

Unfortunately, detailed investigations are still few on the effects of light and, at the current state of our knowledge, it is rather difficult to draw a generalization. To give a few examples from recent literature, Gill et al. (1981) reported that continuous darkness extending even beyond 48 h was best for *Nicotiana plumbaginifolia*. Such is also the case for the legume *Medicago sativa* (Santos et al. 1980; Arcioni et al. 1982). Apparently, Kohlenbach et al. (1982) also found that in *Brassica napus* protoplasts were kept in darkness for as much as 3–4 weeks. However, in marked contrast to these reports, Oelck et al. (1982) have found that of the three legumes investigated, in two, namely, *Trifolium resupinatum* and *Melilotus officinalis,* light was obligatorily required for initiating divisions, though for the third legume, *Arachis hypogaea,* it did not particularly matter either way whether cultures remained in light or dark.

There is some indication that light sensitivity may have a genetic basis. In this connection the earlier work on tobacco of Banks and Evans (1976) is of interest. They found that different species differ considerably in their sensitivity to even moderately low light intensity (700 lx). Though the protoplasts of *Nicotiana tabacum* (in accordance with earlier studies) as well as of *N. sylvestris* showed light tolerance, those of *N. otophora* had an obligatory requirement for darkness. Interestingly, protoplasts from the Fi hybrid of a cross between *N. tabacum* and *N. otophora* were also relatively light insensitive, but overall growth was favoured more in darkness. More recently, Passiatore and Sink (1981) have demonstrated similar differences in some additional species of tobacco. These observations further support the idea that the requirement of light or sensitivity to it may be genetically controlled.

6 General Conclusions, Comments, and Perspectives

Some years ago, Steward and Krikorian (1979) made the interesting comment that the work on free protoplasts has produced more propaganda than substance. Nonetheless, most readers would readily agree that since the first reports on plant regeneration from protoplasts in mosses – *Funaria hygrometrica* and *Physcomitrium eurystomum* (Binding 1964) – and in a higher plant, *Nicotiana tabacum* (Takebe et al. 1971) we have come a long way in exploiting protoplast technology for obtaining entire plants and somatic hybrids. The total number of species in which protoplasts have been successfully regenerated to plantlets or embryos is now nearly a hundred. Yet there is some substance in Steward's statement since even those investigators who have intensively worked on protoplasts have sometimes been wary of the bizarre behaviour of protoplasts. Thus, a glance at Table 1

Table 1. List of plants in which regeneration has been reported[a]

Taxa	Response	Reference
Bryophyta		
Hepaticae		
Marchantia polymorpha	++	Ono et al (1979)
Sphaerocarpos donnellii	++	Wenzel and Schieder (1973)
Musci		
Anoectangium thomsonii	++	Saxena and Rashid (1980)
Funaria hygrometrica	++	Binding (1964, 1966)
Physcomitrella patens	++	Stumm et al. (1975)
Physcomitrium eurystomum	++	Binding (1964, 1966)
Polytrichum juniperinum	++	Gay (1980)
Spermatophyta		
Dicotyledoneae		
Compositae		
Chrysanthemum segetum	+r	Binding et al. (1981)
Cichorium endivia	++	Binding et al. (1981)
C. intybus	++	Binding et al. (1981)
Crepis capillaris	+r	Binding et al. (1981)
Gaillardia grandiflora	++	Binding et al. (1981)
Helianthus annus	+s	Binding et al. (1981)
H. tuberosus	+r	Binding et al. (1981)
Lactuca sativa	++	Berry et al. (1982) Engler and Grogan (1983)
Senecio jacobaea	++	Binding et al. (1981)
S. silvaticus	++	Binding et al. (1981)
S. vernalis	+s	Binding et al. (1981)
S. viscosus	++	Binding et al. (1981)
S. vulgaris	++	Binding and Nehls (1980)
Convolvulaceae		
Pharbitis nil	+r	Messerschmidt (1974)
Cruciferae		
Arabidopsis thaliana	++	Xuan and Menczel (1980)
Brassica campestris	+r	Schenk and Hoffmann (1979)
B. napus	++	Kartha et al. (1974)
B. napus (haploid)	++	Thomas et al. (1976)
B. nigra	+r	Schenk and Hoffmann (1979)
B. oleracea	++	Gatenby and Cocking (1977) Xu et al. (1982)
B. rapa	++	Ulrich et al. (1980)
Sinapis alba	++	Binding et al. (1982)
S. arvensis	++	Binding et al. (1982)
Euphorbiaceae		
Manihot esculenta	++	Shahin and Shepard (1980)
Geraniceae		
Geranium sp.	++	Kameya (1979)
Labiatae		
Majorana hortensis	++	Binding et al. (1982)
Leguminosae		
Clianthus formosus	++	Binding (1984)
Lotus corniculatus	++	Ahuja et al. (1983a)
Medicago coerulea	++	Arcioni et al. (1982)
M. glutinosa	++	Arcioni et al. (1982)
M. sativa	++	Gamborg et al. (1974) Kao and Michayluk (1980)

Table 1 (continued)

Taxa	Response	Reference
Onobrychis viciifolia	++	Ahuja et al. (1983b)
Stylosanthes guyanensis	++	Meijer and Steinbiss (1983)
Trifolium repens	++	Gresshoff (1980)
		Ahuja et al. (1983b)
Trigonella foenum-graecum	+s	Shekhawat and Galston (1983a)
Vicia faba	+r	Binding et al. (1981)
Vigna aconitifolia	++	Shekhawat and Galston (1983b)
V. sinensis	+r, s	Davey et al. (1974)
Linaceae		
Linum usitatissimum	++	Binding et al. (1982)
Ranunculaceae		
Nigella arvensis	++	Binding et al. (1981)
N. damascena	+r	Binding et al. (1981)
N. sativa	+r	Jha and Roy (1979)
Ranunculus sceleratus	++	Dorion et al. (1975)
Resedaceae		
Reseda lutea	++	Binding and Kollmann (1985)
R. luteola	++	Binding et al. (1981)
R. odorata	+s	Binding (pers. commun.)
Rosaceae		
Fragaria ananassa	++	Binding et al. (1982)
Rutaceae		
Citrus aurantium	++	Vardi et al. (1982)
C. limon	++	Vardi et al. (1982)
C. paradisi	++	Vardi et al. (1982)
C. reticulata	++	Vardi et al. (1982)
C. sinensis	++	Vardi et al. (1975)
Scrophulariaceae		
Antirrhinum majus	++	Poirier-Hamon et al. (1974)
Digitalis lanata	++	Xiang-hui Li (1981)
Nemesia strumosa	++	Hess and Leipoldt (1979)
Rehmannia glutinosa	+s	Xu and Davey (1983)
Solanaceae		
Atropa belladonna	++	Gosch et al. (1975b)
Browallia viscosa	++	Power and Berry (1979)
Capsicum annuum	++	Saxena et al. (1981b)
Datura innoxia (diploid and haploid)	++	Schieder (1975)
Datura metel (diploid and haploid)	++	Schieder (1977)
D. meteloides (diploid and haploid)	++	Schieder (1977)
Hyoscyamus albus	+r	Lörz et al. (1979)
H. muticus (diploid and haploid)	++	Lörz et al. (1979)
		Wernicke et al. (1979)
Lycopersicon esculentum	++	Morgan and Cocking (1982)
L. peruvianum	++	Zapata and Sink (1981)
Nicotiana acuminata	++	Bourgin et al. (1979)
N. alata	++	Bourgin and Missonier (1978) (n)
		Bourgin et al. (1979) (2n)
N. debneyi	++	Scowcroft and Larkin (1980)
N. forgetiana	++	Passiatore and Sink (1981)
N. glauca	++	Bourgin et al. (1979)
N. langsdorfii	++	Bourgin et al. (1979)
N. longiflora	++	Bourgin et al. (1979)
N. megalosiphon	++	Shakurov (1982)

Table 1 (continued)

Taxa	Response	Reference
N. neosophila	++	Evans (1979)
N. occidentalis	++	Shakurov (1982)
N. otophora	++	Banks and Evans (1976)
N. paniculata	++	Bourgin et al. (1979)
N. plumbaginifolia (diploid and haploid)	++	Gill et al. (1978) (2n)
		Sidorov et al. (1981) (n)
N. repanda	++	Evans (1979)
N. rustica	++	Gill et al. (1979)
N. sanderae	++	Passiatore and Sink (1981)
N. stocktonii	++	Evans (1979)
N. suaveolens	++	Bourgin et al. (1979)
N. sylvestris	++	Banks and Evans (1976)
		Bourgin et al. (1976)
		Nagy and Maliga (1976)
N. tabacum (diploid and haploid)	++	Nagata and Takebe (1971) (2n)
		Takebe et al. (1971) (2n)
		Ohyama and Nitsch (1972) (n)
N. velutina	++	Shakurov (1982)
Petunia axillaris	++	Power et al. (1976)
P. hybrida (diploid and haploid)	++	Durand et al. (1973) (2n)
		Frearson et al. (1973) (2n)
		Binding (1974) (n)
P. inflata	++	Power et al. (1976)
P. parodii	++	Hayward and Power (1975)
P. parviflora	++	Sink and Power (1977)
P. violaceae	++	Power et al. (1976)
Physalis ixocarpa	+r	Bapat and Schieder (1981)
P. minima	+r	Bapat and Schieder (1981)
Salpiglossis sinuata	++	Boyes et al. (1980)
Solanum brevidens	++	Barsby and Shepard (1983)
		Nelson et al. (1983)
S. chacoense	++	Butenko et al. (1977)
S. dulcamara (diploid and haploid)	++	Binding and Nehls (1977) (2n)
		Binding and Mordhorst (1984)
S. etuberosum	++	Barsby and Shepard (1983)
S. fernandezianum	++	Barsby and Shepard (1983)
S. luteum	++	Binding et al. (1981)
S. melongena	++	Saxena et al. (1981 a)
S. nigrum	++	Nehls (1978)
S. phureja	++	Schumann et al. (1980)
S. tuberosum	++	Upadhya (1975)
		Shepard and Totten (1977)
S. tuberosum (dihaploid)	++	Binding et al. (1978)
S. xanthocarpum	++	Saxena et al. (1982 b)
S. viarum	++	Kowalczyk et al. (1983)
Umbelliferae		
Daucus carota	++	Grambow et al. (1972)
		Kameya and Uchimiya (1972)
Valerianceae		
Kentranthus ruber	+r	Binding et al. (1980)
Monocotyledoneae		
Amaryllidaceae		
Hemerocallis	++	Fitter and Krikorian (1981)

Table 1 (continued)

Taxa	Response	Reference
Gramineae		
Bromus inermis	++	Kao et al. (1973)
Oryza sativa	+r	Deka and Sen (1976)
		Cai et al. (1978)
Panicum maximum	++	Lu et al. (1981)
Pennisetum americanum	++	V. Vasil and I.K. Vasil (1980)
P. purpureum	++	V. Vasil et al. (1983)
Liliaceae		
Asparagus officinalis	++	Bui-Dang-Ha and Mackenzie (1973)

[a] + Organogenesis: r roots, s shoots; ++ plantlets or embryoids

reveals that for some reason, success has been easier to come by in certain plant families, such as the Solanaceae to which nearly half of the total number of species where protoplasts have been successfully regenerated belong. Many other plants have proved extremely retractable to any of the available cultural strategies – unfortunately, this dilemma includes, in particular, cereals and legumes to which most of our important food crops belong.

The mesophyll protoplasts of cereals have been recalcitrant to culture in spite of painstaking efforts in the laboratories of Cocking (Evans et al. 1972), Galston (Brenneman and Galston 1975; Galston et al. 1980), Potrykus (Potrykus et al. 1976; Potrykus 1980), Koblitz (1976), as also the Chinese (Cai et al. 1978) and Hungarian workers (Nemet and Dudits 1977), and others, such as Farmer and Lee (1977) and Thomas (Thomas et al. 1979). The earlier results on *Zea mays* were no different in Vasil's laboratory (V. Vasil and I. K. Vasil 1974). The remark by Potrykus et al. (1976) of their inability of induce growth and division in protoplasts isolated from 75 species and varieties of wheat, barley, rye, oat, and corn, under as many as 80,000 combinations of culture media composition demonstrates the poor regenerative capacity of cereals at least with the current methods.

However, successful culture of cereal protoplasts leading at least to callusing has now been achieved in several plants, such as barley, maize, rice, sorghum, and wheat – a crucial factor being the isolation of protoplasts from selected lines of callus and suspension cultures as illustrated by the work on maize by Potrykus and co-workers (1979a; see also Potrykus 1980). The requirements for regeneration may be even more specific. The investigations of Vasil and co-workers (V. Vasil and I. K. Vasil 1980; Lu et al. 1981; V. Vasil et al. 1983) who obtained adventive embryos and plantlets from protoplasts of *Pennisetum americanum, P. purpureum,* and *Panicum maximum* emphasize in a particularly striking way that the source is an important factor governing their regeneration; according to them, the cell lines derived from immature embryos – which are capable of differentiating plants – are the best source of protoplasts. Indeed, the earlier failures to even induce division in cereal mesophyll protoplasts raised the doubt whether the mesophyll tissue was at all totipotent (Thomas et al. 1980).

The recent work on induction of embryogenesis and differentiation of somatic embryos or plantlets from leaves of cereals and grasses, such as that of Haydu and Vasil (1981), Lu and Vasil (1981), Wernicke and Brettell (1982), and Zamura and Scott (1983) – even though these reports pertain to whole cells – serves to indicate that the conclusion concerning the lack of totopotency in mesophyll cells may be premature and systematic investigations need to be made of the various factors that control the release of totipotency in such tissues. At least in one instance, sustained divisions, though not differentiation, were obtained from protoplasts isolated from wheat leaves (Xiang-hui et al. 1980), which is very encouraging. At any rate, the problem of possible lack of totipotency has shifted from mesophyll cells of leaves, in general, to that of mature leaves. In the light of current knowledge, changes such as transposition of segments of chromosomes or of genes, cannot be ruled out completely even in adult and mature leaves (such changes are now believed to be of common occurrence in microbes as well as eukaryotes), but probably the question now is of how to recall the information for morphogenesis rather than of lack of totipotency *per se*. The problems of differentiation encountered in the cereals have a parallel also in the legumes. Though totipotency of mesophyll cells has not been questioned, even in this group successful regeneration into embryos and plantlets has been achieved in only five genera (see Table 1).

The difficulties in obtaining regeneration can have many facets. To cite one example, Galston and co-workers (Racusen et al. 1977; Kinnersley et al. 1978) found a correlation between the lack of ability of oat protoplasts to divide and the failure to reestablish the negative bioelectric potential on the cell membrane. Normally, the inside of an intact mesophyll cell is electrically negative prior to plasmolysis, but becomes slightly positive when protoplasts are formed. Thus, when potentials are measured with one microelectrode inside the cell or the protoplast and the other in the bathing fluid, the intriguing finding is that whereas in tobacco, the protoplasts of which are able to generate normal cell walls and undergo divisions, the potential returns to a negative value, in protoplasts of oat as well as corn the potential continues to be positive and interestingly both of these fail to regenerate. Although some later work on changes in electric potential is contradictory (e.g. Birskin and Leonard 1979; see also Galun 1981), the merit of the studies of Galston and co-workers lies in focussing attention on a new facet of development of protoplasts and the need to understand the underlying biophysical and biochemical changes inside the protoplasts as well as on the surface of their membranes. As we have mentioned earlier nuclear divisions and cell wall formation may not be obligatorily linked, but generally the two processes proceed side by side, and some relationship between the two is entirely possible.

It is clear that emphasis needs to be given also to various other aspects of the physiology of regeneration of protoplasts. Although encouraged by recent success in regenerating protoplasts from embryogenic tissue, workers have tended to emphasize the choice of donor tissue as compared to such factors as the culture medium, probably the truth is that all aspects, including pretreatments given to the donor tissue, the method of culture, the nutritional milieu, the physical environment, and the genotype, are important. The literature is full of examples where small variations in techniques have resulted in dramatic improvement of plating efficiencies and have often made the difference between success and failure. Even seemingly trivial culture procedures are sometimes vital, such as whether liquid or solid medium has been employed (if liquid medium, whether drop culture or thin-layer method is chosen, or when solid media are used whether a filter paper has been employed), or to give another example, whether protoplasts have been prechilled or not and whether cultures are kept in light or darkness. Nevertheless, equally important is an understanding of the details of the biochemistry of cell wall regeneration as also of control of mitosis for further progress. The monocotyledons have been shown to have a different composition of walls, and one needs to be sure if, in attempts to dissolve the cell walls by currently available enzymes, one does not damage protoplasts irretrievably, and whether the right components are being provided in the culture medium for their regeneration.

Acknowledgements. S.C.M. and N.M. acknowledge the gracious hospitality of Professor A.W. Galston at Yale, where the main frame of this review was prepared and who gave us the benefit of his interest and advice. Sincere thanks are due also to Drs. Shashi Bharal and Praveen Saxena, at Delhi, whose help was also invaluable in preparing the final draft. We are gratful also to Dr. I.K. Vasil and Dr. I. Potrykus for their valuable comments on the manuscript.

Thanks are due also to the Department of Science and Technology of the Government of India for supporting our work and the Jawaharlal Nehru Memorial Fund and the U.S. Educational Foundation in New Delhi for enabling the visit of the senior author to the United States.

References

Abo El-Nil MM, Hildebrandt AC (1976) Cell wall regeneration and colony formation from isolated single geranium protoplasts in microculture. Can J Bot 54:1530–1534

Ahuja PS, Hadiuzzaman S, Davey MR, Cocking EC (1983a) Prolific plant regeneration from protoplast-derived tissue of *Lotus corniculatus* L. (birdsfoot trefoil). Plant Cell Rep 2:101–104

Ahuja PS, Lu DY, Cocking EC, Davey MR (1983b) An assessment of the cultural capabilities of *Trifolium repens* L. (white clover) and *Onobrychis viciifolia* Scop. (Sainfoin) mesophyll protoplasts. Plant Cell Rep 2:269–272

Akerman, KEO, Proudlove MO, Moore AL (1983) Evidence for a Ca^{++} gradient across the plasma membrane of wheat protoplasts. Biochem Biophys Res Commun 113:171–177

Arcioni S, Davey MR, Santos AVP Dos, Cocking EC (1982) Somatic embryogenesis in tissues from mesophyll and cell suspension protoplasts of *Medicago coerulea* and *M. glutinosa*. Z Pflanzenphysiol 106:105–110

Arnold S von, Eriksson T (1976) Factors influencing the growth and division of pea mesophyll protoplasts. Physiol Plant 36:193–196

Arnold S von, Eriksson T (1977) A revised medium for growth of pea mesophyll protoplasts. Physiol Plant 39:257–260

Asamizu T, Nishi A (1980) Regenerated cell wall of carrot protoplasts isolated from suspension-cultured cells. Physiol Plant 48:207–212

Banks MS, Evans PK (1976) A comparison of the isolation and culture of mesophyll protoplasts from several *Nicotiana* species and their hybrids. Plant Sci Lett 7:409–416

Bapat VA, Schieder O (1981) Protoplast culture of several members of the genus *Physalis*. Plant Cell Rep 1:69–70

Barbier M, Dulieu H (1983) Early occurrence of genetic variants in protoplast cultures. Plant Physiol (Bethesda) 29:201–206

Barsby T, Shepard JF (1983) Regeneration of plants from mesophyll protoplasts of *Solanum* species of the *Etuberosa* group. Plant Sci Lett 31:101–105

Berry SF, Lu DY, Pental D, Cocking EC (1982) Regeneration of plants from protoplasts of *Lactuca sativa* L. Z Pflanzenphysiol 108:31–38

Bharal S, Rashid A (1980) Isolation of protoplasts from stem and hypocotyl of the legume *Vigna sinensis* and some factors affecting their regeneration. Protoplasma 102:307–313
Bhatt DP, Fassuliotis G (1981) Plant regeneration from mesophyll protoplasts of eggplant. Z Pflanzenphysiol 104:81–89
Bhojwani SS, Razdan MK (1983) Plant Tissue Culture: Theory and Practice. Elsevier, Amsterdam
Bhojwani SS, Power JB, Cocking EC (1977) Isolation, culture and division of cotton callus protoplasts. Plant Sci Lett 8:85–89
Bidney DL, Shepard JF, Kaleikau E (1983) Regeneration of plants from mesophyll protoplasts of *Brassica oleracea*. Protoplasma 117:89–92
Bilkey PC, Cocking EC (1982) A non-enzymatic method for isolation of protoplasts from callus of *Saintpaulia ionantha* (African violet). Z Pflanzenphysiol 105:285–288
Binding H (1964) Regeneration und Verschmelzung nackter Laubmoosprotoplasten. Z Naturforsch Teil B Anorg Chem Org Chem 19:775
Binding H (1966) Regeneration und Verschmelzung nackter Laubmoosprotoplasten. Z Pflanzenphysiol 55:305–321
Binding H (1974a) Cell cluster formation by leaf protoplasts from axenic cultures of haploid *Petunia hybrida* L. Plant Sci Lett 2:185–188
Binding H (1974b) Regeneration von haploiden und diploiden Pflanzen aus Protoplasten von *Petunia hybrida* L. Z Pflanzenphysiol 74:327–356
Binding H (1984) Aufzucht von Pflanzen aus isolierten Protoplasten und Fusionskörpern. Abstr. In: Mitteilungsband – Kurzfassungen der Beiträge, Botaniker-Tagung in Wien 1984. Inst Bot Univ Wien, p 61
Binding H, Kollmann R (1985, in press) Regeneration of protoplasts. In: Coordination of Agricultural Research. Seminar Proc. Comm Commun, Luxembourg
Binding H, Nehls R (1977) Regeneration of isolated protoplasts to plants in *Solanum dulcamara* L. Z. Pflanzenphysiol 85:279–280
Binding H, Nehls R (1978) Regeneration of isolated protoplasts of *Vicia faba* L. Z Pflanzenphysiol 88:327–332
Binding H, Nehls R (1980) Protoplast regeneration to plants in *Senecio vulgaris* L. Z Pflanzenphysiol 99:183–185
Binding H, Nehls R, Schieder O, Sopory SK, Wenzel G (1978) Regeneration of mesophyll protoplasts isolated from dihaploid clones of *Solanum tuberosum*. Physiol Plant 43:52–54
Binding H, Nehls R, Kock R (1980) Versuche zur Protoplastenregeneration dikotyler Pflanzen unterschiedlicher systematischer Zugehörigkeit. Ber Dtsch Bot Ges 93:667–671
Binding H, Nehls R, Kock R, Finger J, Mordhorst G (1981) Comparative studies on protoplast regeneration in herbaceous species of the Dicotyledoneae class. Z Pflanzenphysiol 101:119–130
Blaschek W, Koehler H, Semler U, Franz G (1982) Molecular weight distribution of cellulose in primary cell walls. Investigations with regenerating protoplasts, suspension cultured cells and mesophyll of tobacco. Planta (Berl) 154:550–555
Bourgin JP, Missonier C (1978) Culture de protoplastes de mésophylle *Nicotiana alata* Link et Otto. Z Pflanzenphysiol 87:55–64
Bourgin JP, Missionier C, Chupeau Y (1976) Culture de protoplastes de mésophylle de *Nicotiana sylvestris* Spegazzini et Comes haplöide et diplöide. CR Acad Sci Paris Ser D 282:1853–1856
Bourgin JP, Chupeau Y, Missionier C (1979) Plant regeneration from mesophyll protoplasts of several *Nicotiana* species. Physiol Plant 45:288–292
Boyes DJ, Sink KC (1981) Regeneration of plants from callus-derived protoplasts of *Salpiglossis*. J Am Soc Hortic Sci 106:42–46
Boyes CJ, Zapata FJ, Sink KC (1980) Isolation, culture and regeneration to plants of callus protoplasts of *Salpiglossis sinuata* L. Z Pflanzenphysiol 99:471–474
Brar DS, Rambold S, Constabel F, Gamborg OL (1980) Isolation, fusion and culture of *Sorghum* and corn protoplasts. Z Pflanzenphysiol 96:269–275
Brenneman FN, Galston AW (1975) Experiments of the cultivation and calli of agriculturally important plants. 1. Oat (*Avena sativa* L.). Biochem Physiol Pflanzen (BPP) 168:453–471
Briskin DP, Leonard RT (1979) Ion transport and isolated protoplasts from tobacco suspension cells. III. Membrane potential. Plant Physiol (Bethesda) 64:959–962

Bui-Dang-Ha D, Mackenzie IA (1973) The division of protoplasts from *Asparagus officinalis* L. and their growth and differentiation. Protoplasma 78:215–221

Burgess J, Linstead PJ (1979) Structure and association of wall fibrils produced by regenerating tobacco protoplasts. Planta (Berl) 146:203–210

Butenko RG, Kuchko AA, Vitenko AA, Aventisov VA (1977) Production and cultivation of protoplasts isolated from the mesophyll of leaves of *Solanum tuberosum* L. and *Solanum chacoense* Bitt. Sov Plant Physiol (Engl Transl Fiziol Rast) 24:540–541

Caboche M (1980) Nutritional requirements of protoplast-derived, haploid tobacco cells grown at low cell densities in liquid medium. Planta (Berl) 149:7–18

Cai Q, Quain Y, Zhou Y, Wu S (1978) A further study on the isolation and culture of rice (*Oryza sativa* L.) protoplasts. Acta Bot Sin 20:97–102

Carlberg I, Glimelius K, Eriksson T (1983) Improved culture ability of potato protoplasts by use of activated charcoal. Plant Cell Rep 2:223–225

Chang T-Y, Senn AA, Pilet PE (1983) Effect of abscisic acid on maize root protoplasts. Z Pflanzenphysiol 110:127–133

Chapman KSR, Hatch MD (1983) Intracellular location of phosphoenol pyruvate carboxykinase and other C-4 photosynthetic enzymes in mesophyll and bundle sheath protoplasts of *Panicum maximum*. Plant Sci Lett 29:145–154

Chupeau Y, Bourgin JP, Missonier C, Dorion N, Morel G (1974) Preparation et culture de protoplastes de divers *Nicotiana*. CR Acad Sci Paris Ser D 278:1564–1568

Cocking EC (1960) A method for isolation of plant protoplasts and vacuoles. Nature (Lond) 187:927–929

Cocking EC (1966) Electron microscopic studies on isolated plant protoplasts. Z Naturforsch Teil B Anorg Chem Org Chem 21:581–584

Cocking EC (1972) Plant cell protoplasts – isolation and development. Annu Rev Plant Physiol 23:29–50

Constabel F, Kirkpatrick JW, Gamborg OL (1973) Callus formation from mesophyll protoplasts of *Pisum sativum*. Can J Bot 51:2105–2106

Cornel D, Grignon C, Rona JP, Heller R (1983) Measurement of intracellular potassium activity in protoplasts of *Acer pseudoplatanus:* origin of their electropositivity. Physiol Plant 57:203–209

Crepy L, Chupeau M-C, Chupeau Y (1982) The isolation and culture of leaf protoplasts of *Cichorium intybus* and their regeneration into plants. Z Pflanzenphysiol 107:123–131

Davey MR, Bush E, Power JB (1974) Cultural studies of a dividing legume leaf protoplast system. Plant Sci Lett 3:127–133

Davey MR, Mathias RJ (1979) Close-packing of plasma membrane particles during wall regeneration by isolated higher plant protoplasts – fact or artifact? Protoplasma 100:85–99

Deka PC, Sen SK (1976) Differentiation in calli originated from isolated protoplasts of rice (*Oryza sativa* L.) through plating technique. Mol Gen Genet 145:239–243

Donn G (1978) Cell division and callus regeneration from leaf protoplasts of *Vicia narbonensis*. Z Pflanzenphysiol 86:65–75

Dorion N, Chupeau Y, Bourgin JP (1975) Isolation, culture and regeneration into plants of *Ranunculus sceleratus* L. leaf protoplasts. Plant Sci Lett 5:325–331

Dudits K, Kao KN, Constabel F, Gamborg OL (1976) Embryogenesis and formation of tetraploid and hexaploid plants from carrot protoplasts. Can J Bot 5:1063–1067

Durand J, Potrykus I, Donn G (1973) Plantes issues de protoplastes de *Petunia*. Z Pflanzenphysiol 69:26–34

Engler DE, Grogan RG (1983) Isolation, culture and regeneration of lettuce leaf mesophyll protoplasts. Plant Sci Lett 28:223–229

Enzmann-Becker G (1973) Plating efficiency of protoplasts of tobacco in different light conditions. Z Naturforsch Sect C Biosci 28:470–471

Eriksson T, Jonasson K (1969) Nuclear division in isolated protoplasts from cells of higher plants grown *in vitro*. Planta (Berl) 89:85–89

Evans DA (1979) Chromosome stability of plants regenerated from mesophyll protoplasts of *Nicotiana* species. Z Pflanzenphysiol 95:459–463

Evans PK, Keates AG, Cocking EC (1972) Isolation of protoplasts from cereal leaves. Planta (Berl) 104:178–181

Evola SV, Earle ED, Chaleff RS (1983) The use of genetic markers selected in vitro for the isolation and genetic verification of intraspecific somatic hybrids of *Nicotiana tabacum* L. Mol Gen Genet 189:441–446

Farmer I, Lee PE (1977) Culture of protoplasts derived from Ramsey durum wheat. Plant Sci Lett 10:141–145

Fitter MS, Krikorian AD (1981) Recovery of totipotent cells and plantlet production from day lily protoplasts. Ann Bot (Lond) 48:591–597

Fowke LC, Bech-Hansen CW, Constabel F, Gamborg OL (1974) A comparative study on the ultrastructure of cultured cells and protoplasts of soybean during cell division. Protoplasma 81:189–203

Frearson EM, Power JB, Cocking EC (1973) The isolation, culture and regeneration of *Petunia* leaf protoplasts. Dev Biol 33: 130–137

Furner IJ, King J, Gamborg OL (1978) Plant regeneration from protoplasts isolated from a predominantly haploid suspension culture of *Datura innoxia* Mill. Plant Sci Lett 11:169–176

Galbraith DW (1981) Microfluorimetric quantitation of cellulose biosynthesis by plant protoplasts using Calcofluor White. Physiol Plant 53:111–116

Galbraith DW, Shields BA (1982) The effects of inhibitors of cell wall synthesis on tobacco protoplast development. Physiol Plant 55:25–30

Galston AW, Kaur-Sawhney R, Altman A, Flores H (1980) Polyamines, macromolecular synthesis and the problem of cereal protoplast regeneration. In: Ferenczy L, Farkas GL (eds) Advances in Protoplast Research. Pergamon, Oxford, pp 485–497

Galun E (1981) Plant protoplasts as physiological tools. Annu Rev Plant Physiol 32:237–266

Gamborg OL, Miller RA, Ojima K (1968) Nutrient requirements of suspension cultures of soybean root cells. Exp Cell Res 50:151–158

Gamborg OL, Miller RA (1973) Isolation, culture and uses of plant protoplasts. Can J Bot 51:1795–1799

Gamborg OL, Constabel F, Fowke L, Kao KN, Ohyama K, Kartha K, Pelcher L (1974) Protoplast and cell culture methods in somatic hybridization in higher plants. Can J Genet Cytol 16:737–750

Gamborg OL, Shyluk JP, Kartha KK (1975) Factors affecting the isolation and callus formation in protoplasts from the shoot apices of *Pisum sativum* L. Plant Sci Lett 4:285–292

Gamborg OL, Davis BP, Stahlhut RW (1983) Cell division and differentiation in protoplasts from cell cultures of *Glycine* species and leaf tissue of soybean. Plant Cell Rep 2:213–215

Gatenby AA, Cocking EC (1977) Callus formation from protoplasts of marrow stem kale. Plant Sci Lett 8:275–280

Gay L (1980) The development of leafy gametophytes from isolated protoplasts of *Polytrichum juniperinum* Willd. Z Pflanzenphysiol 79:33–39

Gill R, Rashid A, Maheshwari SC (1978) Regeneration of plants from mesophyll protoplasts of *Nicotiana plumbaginifolia* Viv. Protoplasma 96:375–379

Gill R, Rashid A, Maheshwari SC (1979) Isolation of mesophyll protoplasts of *Nicotiana rustica* and their regeneration into plants flowering in vitro. Physiol Plant 47:7–10

Gill R, Rashid A, Maheshwari SC (1981) Dark requirement for cell regeneration and colony formation by mesophyll protoplasts of *Nicotiana plumbaginifolia* Viv. Protoplasma 106:351–354

Gleba YY (1978) Microdroplet culture: Tobacco plants from single mesophyll protoplasts. Naturwissenschaften 65:158–159

Gosch G, Bajaj YPS, Reinert J (1975a) Isolation, culture, and fusion studies on protoplasts from different species. Protoplasma 85:327–336

Gosch G, Bajaj YPS, Reinert J (1975b) Isolation, culture, and induction of embryogenesis in protoplasts from cell-suspensions of *Atropa belladonna*. Protoplasma 86:405–410

Grambow HJ, Kao KN, Miller RA, Gamborg OL (1972) Cell divisions and plant development from protoplasts of carrot cell suspension cultures. Planta (Berl) 103:348–355

Gresshoff PM (1980) In vitro culture of white clover: callus, suspension, protoplast culture, and plant regeneration. Bot Gaz 141:157–164

Hahne G, Herth W, Hoffmann F (1983) Wall formation and cell division in fluorescence-labelled plant protoplasts. Protoplasma 115:217–221

Halim H, Pearce RS (1980) An electrophoresis method for bulk manipulation of isolated protoplasts from higher plants. Biochem Physiol Pflanz (BPP) 175:123–129

Hanke DE, Northcote DH (1974) Cell wall formation by soybean callus protoplasts. J Cell Sci 14:29–50

Harms CT, Potrykus I (1978) Enrichment for heterokaryocytes by the use of iso-osmotic density gradients after plant protoplast fusion. Theor Appl Genet 53:49–55

Harms CT, Lörz H, Potrykus I (1979) Multiple-drop-array (MDA) technique for the large-scale testing of culture media variations in hanging microdrop cultures of single cell systems. II. Determination of phytohormone combinations for optimal division response in *Nicotiana tabacum* protoplast cultures. Plant Sci Lett 14:237–244

Hasezawa S, Nagata T, Syono K (1981) Transformation of *Vinca* protoplasts mediated by *Agrobacterium* spheroplasts. Mol Gen Genet 182:206–210

Haydu Z, Vasil IK (1981) Somatic embryogenesis and plant regeneration from leaf tissues and anthers of *Pennisetum purpureum* Schum. Theor Appl Genet 59:269–273

Hayward C, Power JB (1975) Plant production from leaf protoplasts of *Petunia parodii*. Plant Sci Lett 4:407–410

Heber U (1982) Photosynthesis under osmotic stress: effect of high solute concentrations on the permeability properties of the chloroplast envelope and on activity of stroma enzymes. Planta (Berl) 153:423–429

Herth W, Meyer Y (1977) Ultrastructural and chemical analysis of the wall fibrils synthesized by tobacco mesophyll protoplasts. Biol Cell 30:33–40

Hess D, Leipoldt G (1979) Regeneration of roots and shoots from isolated mesophyll protoplasts of *Nemesia strumosa*. Biochem Physiol Pflanz (BPP) 174:411–417

Hooley R (1982) Protoplasts isolated from aleurone layers of wild oat (*Avena fatua* L.) exhibit the classic response to gibberellic acid. Planta (Berl) 154:29–40

Hughes BG, White FG, Smith MA (1976) Scanning electron microscopy of barley protoplasts. Protoplasma 90:399–405

Huhtinen O, Honkanen J, Simola LK (1982) Ornithine- and putrescine-supported divisions and cell colony formation in leaf protoplasts of alders (*Alnus glutinosa* and *A. incana*). Plant Sci Lett 28:3–9

Ishii S, Mogi Y (1983) Identification of enzymes that are effective for isolating protoplasts from grass leaves. Plant Physiol (Bethesda) 72:641–644

Izhar S, Power JB (1977) Genetical studies with *Petunia* leaf protoplasts. I. Genetic variation to specific growth hormones and possible genetic control on stages of protoplast development in culture. Plant Sci Lett 8:375–383

Jha TB, Roy SC (1979) Rhizogenesis from *Nigella sativa* protoplasts. Protoplasma 101:139–142

Jia SR (1982) Factors affecting the division frequency of mesophyll protoplasts. Can J Bot 60:2192–2196

Johnson LB, Stuteville DL, Higgens RK, Douglas HL (1982) Pectolyase Y-23 for isolating mesophyll protoplasts from several *Medicago* species. Plant Sci Lett 26:133–137

Kaiser G, Heber U (1983) Photosynthesis of leaf cell protoplasts and permeability of plasmalemma to some solute. Planta (Berl) 157:462–470

Kameya T (1975) Culture of protoplasts from chimeral plant tissue of nature. Jpn J Genet 50:417–420

Kameya T, Uchimiya H (1972) Embryoids derived from isolated protoplasts of carrot. Planta (Berl) 103:356–360

Kao KN, Keller WA, Miller RA (1970) Cell division in newly formed cells from protoplasts of soybean. Exp Cell Res 62:338–340

Kao KN, Gamborg OL, Miller RA, Keller WA (1971) Cell divisions in cells regenerated from protoplasts of soybean and *Haplopappus gracilis*. Nature New Biol 232:124

Kao KN, Gamborg OL, Michayluk MR, Keller WA, Miller RA (1973) The effects of sugars and inorganic salts on cell regeneration and sustained division in plant protoplasts. In: Protoplastes et fusion de cellules somatiques vegetales. Colloq Int Cent Natl Rech Sci 212:207–213

Kao KN, Michayluk MR (1975) Nutritional requirements for growth of *Vicia hajastana* cells and protoplasts at a very low population density in liquid media. Planta (Berl) 126:105–110

Kao KN, Michayluk MR (1980) Plant regeneration from mesophyll protoplasts of alfalfa. Z Pflanzenphysiol 96:135–141

Kartha KK, Michayluk MR, Kao KN, Gamborg OL, Constabel F (1974) Callus formation and plant regeneration form mesophyll protoplasts of rape plants (*Brassica napus* L. cv. Zephyr). Plant Sci Lett 3:265–271

Kaur-Sawhney R, Flores HE, Galston AW (1980) Polyamine-induced DNA synthesis and mitosis in oat leaf protoplasts. Plant Physiol (Bethesda) 65:368–371

Kim IS, Song PS (1981) Binding of phytochrome to liposomes and protoplasts. Biochemistry 20:5482–5489

Kinnersley AM, Racusen RH, Galston AW (1978) A comparison of regenerated cell walls in tobacco and cereal protoplasts. Planta (Berl) 139:155–158

Kirby EG, Cheng T-Y (1979) Colony formation from protoplasts derived from Douglas fir cotyledons. Plant Sci Lett 14:145–154

Klein AS, Montezinos D, Delmer DP (1981) Cellulose and 1,3-glucan synthesis during the early stages of wall regeneration in soybean protoplasts. Planta (Berl) 152:105–114

Klercker I (1892) Eine Methode zur Isolierung lebender Protoplasten. Öfvers Vet-Akad Förhdl 9:463–474

Koblitz H (1976) Isolierung und Kultivierung von Protoplasten aus Calluskulturen der Gerste. Biochem Physiol Pflanz (BPP) 170:287–293

Kohlenbach HW, Bohnke E (1975) Isolation and culture of mesophyll protoplasts of *Hyoscyamus niger* L. var. *annuus* Sims. Experientia (Basel) 31:1281–1282

Kohlenbach HW, Wenzel C, Hoffmann F (1982) Regeneration of *Brassica napus* plantlets in cultures from isolated protoplasts of haploid stem embryos as compared with leaf protoplasts. Z Pflanzenphysiol 105:131–142

Koop H-U, Weber G, Schweiger H-G (1983) Individual culture of selected single cells and protoplasts of higher plants in microdroplets of defined media. Z Pflanzenphysiol 112:21–34

Kowalczyk TP, Mackenzie IA, Cocking EC (1983) Plant regeneration from organ explants and protoplasts of the medicinal plant *Solanum khasianum* CB Clarke var. *chatterjeeanum* Sengupta (Syn. *Solanum viarum* Dunal). Z Pflanzenphysiol 111:55–68

Krikorian AD (1982) Cloning higher plants from aseptically cultured tissues and cells. Biol Rev Camb Philos Soc 57:151–218

Lloyd CW, Slabas AR, Powell AJ, Lowe SB (1980) Microtubules, protoplasts and plant cell shape, an immunofluorescent study. Planta (Berl) 147:500–506

Lörz H, Potrykus I, Thomas E (1977) Somatic embryogenesis from tobacco protoplasts. Naturwissenschaften 64:439–440

Lörz H, Wernicke W, Potrykus I (1979) Culture and plant regeneration of *Hyoscyamus* protoplasts. Planta Med 36:21–29

Lu CY, Vasil IK (1981) Somatic embryogenesis and plant regeneration in tissue cultures of *Panicum maximum* Jacq. Am J Bot 69:77–81

Lu CY, Vasil V, Vasil IK (1981) Isolation and culture of protoplasts of *Panicum maximum* Jacq. (Guinea grass): somatic embryogenesis and plantlet formation. Z Pflanzenphysiol 104:311–318

Lu DY, Cooper-Bland S, Pental D, Cocking EC, Davey MC (1983) Isolation and sustained division of protoplasts from cotyledons of seedlings and immature seeds of *Glycine max* L. Z Pflanzenphysiol 111:389–394

Lu DY, Davey MR, Cocking EC (1983) A comparison of the cultural behaviour of protoplasts from leaves, cotyledons and roots of *Medicago sativa*. Plant Sci Lett 31:87–89

Mackenzie IA, Bui-Dang-Ha D, Davey MR (1973) Some aspects of the isolation, fine structure and growth of protoplasts from *Asparagus officinalis* L. In: Protoplastes et fusion de cellules somatiques vegetales. Colloq Int Cent Natl Rech Sci 212:291–299

Maheshwari SC, Rashid A, Tyagi AK (1982) Haploids from pollen grains – retrospect and prospect. Am J Bot 69:865–879

Meijer EGM, Steinbiss H-H (1983) Plantlet regeneration from suspension and protoplast cultures of the tropical pasture legume *Stylosanthes guyanensis* (Aubl.) Sw. Ann Bot (Lond) 52:305–310

Menczel L, Lazar C, Maliga P (1978) Isolation of somatic hybrids by cloning *Nicotiana* heterokaryons in nurse cultures. Planta (Berl) 143:29–32

Messerschmidt M (1974) Callus formation and differentiation of isolated protoplasts from cotyledons of *Pharbitis nil*. Z Pflanzenphysiol 74:175–178

Meyer Y (1974) Isolation and culture of tobacco mesophyll protoplasts using a saline medium. Protoplasma 81:363–372

Meyer Y, Abel WO (1975) Budding and cleavage division of tobacco mesophyll protoplasts in relation to pseudo-wall and wall formation. Planta (Berl) 125:1–13

Meyer Y, Cooke R (1979) Time course of hormonal control of the first mitosis in tobacco mesophyll protoplasts cultivated in vitro. Planta (Berl) 147:181–185

Meyer Y, Herth W (1978) Chemical inhibition of cell wall formation and cytokinesis, but not of nuclear division in protoplasts of *Nicotiana tabacum* L. cultivated *in vitro*. Planta (Berl) 142:253–262

Michayluk MR, Kao KN (1975) A comparative study of sugars and sugar alcohols on cell regeneration and sustained cell division in plant protoplasts. Z Pflanzenphysiol 75:181–185

Morgan A, Cocking EC (1982) Plant regeneration from protoplasts of *Lycopersicon esculentum* Mill. Z Pflanzenphysiol 106:97–104

Mühlbach HP (1980) Different regeneration potentials of mesophyll protoplasts from cultivated and a wild species of tomato. Planta (Berl) 148:89–96

Mühlbach HP, Thiele H (1981) Response to chilling of tomato mesophyll protoplasts. Planta (Berl) 151:399–401

Muller JF, Missionier C, Caboche M (1983) Low density growth of cells derived from *Nicotiana* and *Petunia* protoplasts: influence of the source of protoplasts and comparison of growth-promoting activity of various auxins. Physiol Plant 57:35–41

Murashige T, Skoog F (1962) A revised medium for rapid growth and bio-assays with tobacco tissue cultures. Physiol Plant 15:473–497

Nagata T, Ishii S (1979) A rapid method for isolation of mesophyll protoplasts. Can J Bot 57:1820–1823

Nagata T, Takebe I (1970) Cell wall regeneration and cell division in isolated tobacco mesophyll protoplasts. Planta (Berl) 92:301–308

Nagata T, Takebe I (1971) Plating of isolated tobacco mesophyll protoplasts on agar medium. Planta (Berl) 99:12–20

Nagy JI, Maliga P (1976) Callus induction and plant regeneration from mesophyll protoplasts of *Nicotiana sylvestris*. Z Pflanzenphysiol 78:453–455

Nebiolo CM, Kaczmarczyk WJ, Ulrich V (1983) Manifestation of hybrid vigor in RNA synthesis parameters by corn seedling protoplasts in the presence and absence of gibberellic acid. Plant Sci Lett 28:195–206

Negrutiu I, Mousseau J (1980) Protoplast culture from in vitro grown plants of *Nicotiana sylvestris* Spegg. and Comes. Z Pflanzenphysiol 100:373–376

Nehls R (1978) Isolation and regeneration of protoplasts from *Solanum nigrum* L. Plant Sci Lett 12:183–187

Nelson RS, Creissem GP, Bright SWJ (1983) Plant regeneration from protoplasts of *Solanum brevidens*. Plant Sci Lett 30:355–362

Nemet G, Dudits D (1977) Potentials of protoplasts, cell and tissue culture in cereal research. In: Novak FJ (ed) Use of tissue culture in plant breeding. Czechoslovak Acad Sci, Prague, pp 145–163

Norman HA, Black M, Chapman JM (1983) The induction of sensitivity to gibberellin in aleurone tissue of developing wheat grain: sensitization of isolated protoplasts. Planta (Berl) 158:264–271

Oelck MM, Bapat VA, Schieder O (1982) Protoplast culture of three legumes: *Arachis hypogaea, Melilotus officinalis, Trifolium resupinatum*. Z Pflanzenphysiol 106:173–177

Ohyama K, Nitsch JP (1972) Flowering haploid plants obtained from protoplasts of tobacco leaves. Plant Cell Physiol 13:229–236

Ono K, Ohyama K, Gamborg OL (1979) Regeneration of the liverwort *Marchantia polymorpha* L. from protoplasts isolated from cell suspension culture. Plant Sci Lett 14:225–229

Passiatore JE, Sink KC (1981) Plant regeneration from leaf mesophyll protoplasts of selected ornamental *Nicotiana* species. J Soc Hortic Sci 106:779–803

Patnaik G, Wilson D, Cocking EC (1981) Importance of enzyme purification of increased plating efficiency and plant regeneration from single protoplasts of *Petunia parodii*. Z Pflanzenphysiol 102:199–205

Pelcher LE, Gamborg OL, Kao KN (1974) Bean mesophyll protoplasts: production, culture and callus formation. Plant Sci Lett 3:107–111

Poirier-Hamon S, Rao PS, Harada H (1974) Culture of mesophyll protoplasts and stem segments of *Antirrhinum maius* (Snapdragon): Growth and organization of embryoids. J Exp Bot 25:752–760

Potrykus I (1980) The old problem of protoplast culture: cereals. In: Ferenczy L, Farkas GL (eds) Advances in Protoplast Research. Pergamon, Oxford, pp 243–254

Potrykus I, Durand J (1972) Callus formation from single protoplasts of *Petunia*. Nature New Biol 237:286–287

Potrykus I, Harms CT, Lörz H (1976) Problems in culturing cereal protoplasts. In: Dudits D, Farkas GL, Maliga P (eds) Cell Genetics in Higher Plants. Publishing House of the Hungarian Academy of Sciences, Budapest, pp 129–140

Potrykus I, Harms CT, Lörz H, Thomas E (1977) Callus formation from stem protoplasts of corn (*Zea mays* L.) Mol Gen Genet 156:347–350

Potrykus I, Harms CT, Lörz H (1979a) Callus formation from cell culture protoplasts of corn (*Zea mays* L.). Theor Appl Genet 54:209–214

Potrykus I, Harms CT, Lörz H (1979b) Multiple-drop-array (MDA) technique for the large-scale testing of culture media variations in hanging microdrop cultures of single cell systems. I. The technique. Plant Sci Lett 14:231–235

Power JB, Berry SF (1979) Plant regeneration from protoplasts of *Browallia viscosa*. Z Pflanzenphysiol 94:469–471

Power JB, Frearson EM, George D, Evans PK, Berry SF, Hayward C, Cocking EC (1976) The isolation, culture and regeneration of leaf protoplasts in the genus *Petunia*. Plant Sci Lett 7:1–55

Racusen RH, Kinnersley AM, Galston AW (1977) Osmotically induced changes in electrical properties of plant protoplast membranes. Science (Wash DC) 198:405–407

Rahat M, Reinhold L (1983) Rb^+ uptake by isolated pea mesophyll protoplasts in light and darkness. Physiol Plant 59:83–90

Rao PS (1982) Protoplast culture. In: Johri BM (ed) Experimental Embryology of Vascular Plants. Springer, Berlin Heidelberg New York, pp 231–262

Raveh D, Galun E (1975) Rapid regeneration of plants from tobacco protoplasts plated at low densities. Z Pflanzenphysiol 76:76–79

Raveh D, Huberman E, Galun E (1973) In vitro culture of tobacco protoplasts: use of feeder techniques to support division of cells plated at low densities. In Vitro (Rockville) 9:216–222

Roscoe DH, Bell GM (1981) Use of a pH indicator in protoplast culture medium. Plant Sci Lett 21:275–279

Ruesink AW, Thimann KV (1965) Protoplasts from the *Avena* coleoptile. Proc Natl Acad Sci USA 54:56–64

Santos AVF Dos, Outka DE, Cocking EC, Davey MR (1980) Organogenesis and somatic embryogenesis in tissues derived from leaf protoplasts and leaf explants of *Medicago sativa*. Z Pflanzenphysiol 99:261–270

Saxena PK, Rashid A (1980) Development of gametophores from isolated protoplasts of the moss *Anoectangium thomsonii* Mitt. Protoplasma 103:401–404

Saxena PK, Rashid A (1981) High frequency regeneration of *Funaria hygrometrica* protoplasts isolated from low calcium protonemal suspension. Plant Sci Lett 23:117–122

Saxena PK, Gill R, Rashid A, Maheshwari SC (1981a) Plantlet formation from isolated protoplasts of *Solanum melongena* L. Protoplasma 106:355–359

Saxena PK, Gill R, Rashid A, Maheshwari SC (1981b) Isolation and culture of protoplasts of *Capsicum annuum* L. and their regeneration into plants flowering in vitro. Protoplasma 108:357–360

Saxena PK, Gill R, Rashid A, Maheshwari SC (1982a) Colony formation by cotyledonary protoplasts of *Cyamopsis tetragonoloba* L. Z Pflanzenphysiol 106:277–280

Saxena PK, Gill R, Rashid A, Maheshwari SC (1982b) Plantlets from mesophyll protoplasts of *Solanum xanthocarpum*. Plant Cell Rep 1:219–220

Schenk HR, Hoffmann F (1979) Callus and root regeneration from mesophyll protoplasts of basic *Brassica* species: *Brassica campestris, B. oleracea* and *B. nigra*. Z Pflanzenzuecht 82:354–360

Schieder O (1975) Regeneration von haploiden und diploiden *Datura innoxia* Mill. Mesophyll Protoplasten zu Pflanzen. Z Pflanzenphysiol 76:462–466

Schieder O (1977) Attempts in regeneration of mesophyll protoplasts of haploid and diploid wild type lines, and those of chlorophyll-deficient strains from different Solanaceae. Z Pflanzenphysiol 84:275–281

Schilde-Rentschler L (1977) Role of the cell wall in the ability of tobacco protoplasts to form callus. Planta (Berl) 135:177–181

Schumann U, Opatrny Z, Koblitz H (1980) Plant recovery from long term callus cultures and from suspension culture derived protoplasts of *Solanum phureja*. Biochem Physiol Pflanz (BPP) 175:670–675

Scowcroft WR, Larkin PJ (1980) Isolation, culture and plant regeneration from protoplasts of *Nicotiana debneyi*. Aust J Plant Physiol 7:635–644

Shahin EA, Shepard JF (1980) Cassava mesophyll protoplasts: Isolation, proliferation, and shoot formation. Plant Sci Lett 17:459–465

Shakurov MI (1982) Cultivation of isolated protoplasts of different species of the genus *Nicotiana*. Sov Plant Physiol (Engl Transl Fiziol Rast) 29:132–140

Shekhawat NS, Galston AW (1983a) Mesophyll protoplasts of fenugreek (*Trigonella foenum-graecum*): isolation, culture and shoot regeneration. Plant Cell Rep 2:119–121

Shekhawat NS, Galston AW (1983b) Isolation, culture and regeneration of mothbean *Vigna aconitifolia* leaf protoplasts. Plant Sci Lett 32:43–51

Shepard JF (1980) Mutant selection and plant regeneration from potato mesophyll protoplasts. In: Rubenstein I, Gengenbach B, Phillips RL, Green CE (eds) Emergent techniques for the genetic improvement of crops. Univ Minnesota Press, Minneapolis, pp 185–219

Shepard JF, Totten RE (1975) Isolation and regeneration of tobacco mesophyll cell protoplasts under low osmotic conditions. Plant Physiol (Bethesda) 55:689–694

Shepard JF, Totten RE (1977) Mesophyll cell protoplasts of potato – isolation, proliferation, and plant regeneration. Plant Physiol (Bethesda) 60:313–316

Shillito RD, Paszkowski J, Potrykus I (1983) Agarose plating and a bead type culture technique enable and stimulate development of protoplast-derived colonies in a number of plant species. Plant Cell Rep 2:244–247

Sidorov V, Menczel L, Maliga P (1981) Isoleucine-requiring *Nicotiana* plant deficient in threonine deaminase. Nature (Lond) 294:87–88

Simmonds JA, Simmonds DH, Cumming BG (1979) Isolation and cultivation of protoplasts from morphogenetic callus cultures of *Lilium*. Can J Bot 57:512–516

Sink KC, Power JB (1977) The isolation, culture and regeneration of leaf protoplasts of *Petunia parviflora* Juss. Plant Sci Lett 10:335–340

Steward FC, Krikorian AD (1979) Problems and potentialities of cultured plant cells in retrospect and prospect. In: Sharp WR, Larsen FO, Paddock EF, Raghavan V (eds) Plant cell tissue culture, principles and applications. Ohio State Univ Press, Columbus, pp 221–262

Stumm I, Meyer Y, Abel WO (1975) Regeneration of the moss *Physcomitrella patens* Hedw. from isolated protoplasts. Plant Sci Lett 5:113–118

Takebe I, Otsuki Y (1973) Fine structure of isolated mesophyll protoplasts of tobacco. Planta (Berl) 113:21–27

Takebe I, Labib G, Melchers G (1971) Regeneration of whole plants from isolated mesophyll protoplasts of tobacco. Naturwissenschaften 58:318–320

Takeuchi Y, Komamine A (1978) Composition of the cell wall formed by protoplasts isolated from cell suspension cultures of *Vinca rosea*. Planta (Berl) 140:227–232

Takeuchi Y, Komamine A (1982) Effects of culture conditions on cell division and composition of regenerated cell walls in *Vinca rosea* protoplasts. Plant Cell Physiol 23:249–255

Thomas E, Hoffmann F, Potrykus I, Wenzel G (1976) Protoplast regeneration and stem embryogenesis of haploid androgenetic rape. Mol Gen Genet 145:245–247

Thomas E, King PJ, Potrykus I (1979) Improvement of crop plants via single cells in vitro – an assessment. Z Pflanzenzuecht 82:1–30

Thomas E, Brettell R, Wernicke W (1980) Problems in plant regeneration from protoplasts of important crops. In: Ferenczy L, Farkas GL (eds) Advances in protoplasts research. Pergamon, Oxford, pp 269–274

Thomas E (1981) Plant regeneration from shoot culture-derived protoplasts of tetraploid potato (*Solanum tuberosum* cv. Maris Bard). Plant Sci Lett 23:81–88

Thomas E, Bright SWJ, Franklin J, Lancaster VA, Miflin MJ, Gibson R (1982) Variation amongst protoplast-derived potato plants (*Solanum tuberosum* cv. Maris Bard). Theor Appl Genet 72:65–68

Uchimiya H, Murashige T (1976) Influence of the nutrient medium on the recovery of dividing cells from tobacco protoplasts. Plant Physiol (Bethesda) 57:424–429

Ulrich TH, Chowdhury JB, Widholm JM (1980) Callus and root formation from mesophyll protoplasts of *Brassica rapa*. Plant Sci Lett 19:347–354

Upadhya MD (1975) Isolation and culture of mesophyll protoplasts of potato (*Solanum tuberosum* L.). Potato Res 18:438–445

Vardi A, Spiegel-Roy P, Galun E (1975) *Citrus* cell culture: isolation of protoplasts, plating densities, effect of mutagens and regeneration of embryos. Plant Sci Lett 4:231–236

Vardi A, Spiegel-Roy P, Galun E (1982) Plant regeneration from *Citrus* protoplasts: variability in methodological requirements among cultivars and species. Theor Appl Genet 62:171–176

Vasil IK (1982) Plant cell culture and somatic cell genetics of cereals and grasses. In: Vasil IK, Scowcroft WR, Frey KJ (eds) Plant improvement and somatic cell genetics. Academic, New York, pp 179–203

Vasil IK (ed) (1984) Cell culture and somatic cell genetics of plants. Vol 1. Laboratory procedures and their applications. Academic, New York

Vasil IK, Vasil V (1980) Isolation and culture of protoplasts. Int Rev Cytol Suppl 11B:1–19

Vasil V, Vasil IK (1974) Regeneration of tobacco and *Petunia* plants from protoplasts and culture of corn protoplasts. In Vitro (Rockville) 10:83–96

Vasil V, Vasil IK (1979) Isolation and culture of cereal protoplasts. I. Callus formation from pearl millet (*Pennisetum americanum*) protoplasts. Z Pflanzenphysiol 92:379–384

Vasil V, Vasil IK (1980) Isolation and culture of cereal protoplasts. II. Embryogenesis and plantlet formation from protoplasts of *Pennisetum americanum*. Theor Appl Genet 56:97–99

Vasil V, Wang D-Y, Vasil IK (1983) Plant regeneration from protoplasts of napier grass (*Pennisetum purpureum* Schum.). Z Pflanzenphysiol 111:233–239

Wallin A, Eriksson T (1973) Protoplast culture from cell suspensions of *Daucus carota*. Physiol Plant 28:33–39

Wenzel G (1973) Isolation of leaf protoplasts from haploid plants of *Petunia*, rape and rye. Z Pflanzenzuecht 69:58–61

Wenzel G, Schieder O (1973) Regeneration of isolated protoplasts from nicotinic acid-deficient mutants of the liverwort *Sphaerocarpos donnellii* Aust. Plant Sci Lett 1:421–423

Wernicke W, Brettell RIS (1982) Morphogenesis from cultured leaf tissue of *Sorghum bicolor* – culture initiation. Protoplasma 111:19–27

Wernicke W, Thomas E (1980) Studies on morphogenesis from isolated plant protoplasts: shoot formation from mesophyll protoplast of *Hyoscyamus muticus* and *Nicotiana tabacum*. Plant Sci Lett 17:401–407

Wernicke W, Lörz H, Thomas E (1979) Plant regeneration from leaf protoplasts of haploid *Hyoscyamus muticus* L. produced via anther culture. Plant Sci Lett 15:239–249

Willison JHM, Cocking EC (1975) Microfibril synthesis at the surface of isolated tobacco mesophyll protoplasts, a freeze-etch study. Protoplasma 84:147–159

Willison JHM, Grout BWW (1978) Further observations on cell-wall formation around isolated protoplasts of tobacco and tomato. Planta (Berl) 140:53–58

Xiang-hui Li (1981) Plantlet regeneration from mesophyll protoplasts of *Digitalis lanata* Ehrh. Theor Appl Genet 60:345–347

Xiang-hui Li, Yan Qui-sheng, Huang Mei-juan, Sun Yung-ru, Li Wen-bin (1980) Division of cells regenerated from mesophyll protoplasts of wheat (*Triticum aestivum* L.) In: Ferenczy L, Farkas GL (eds) Advances in protoplast research. Pergamon, Oxford, pp 261–267

Xu X-H, Davey MR (1983) Shoot regeneration from mesophyll protoplasts and leaf explants of *Rehmannia glutinosa*. Plant Cell Rep 2:55–57

Xu Z-H, Davey MR, Cocking EC (1981) Isolation and sustained division of *Phaseolus aureus* (mung bean) root protoplasts. Z Pflanzenphysiol 104:289–298

Xu Z-H, Davey MR, Cocking EC (1982) Plant regeneration from root protoplasts of *Brassica*. Plant Sci Lett 24:117–121

Xuan LT, Menczel L (1980) Improved protoplast culture and plant regeneration from protoplast-derived callus in *Arabidopsis thaliana*. Z Pflanzenphysiol 96:77–80

Zamura AB, Scott KI (1983) Callus formation and plant regeneration from wheat leaves. Plant Sci Lett 29:183–189

Zapata FJ, Sink KC (1981) Somatic embryogenesis from *Lycopersicon peruvianum* leaf mesophyll protoplasts. Theor Appl Genet 59:265–268

Zapata FJ, Evans PK, Power JB, Cocking EC (1977) The effect of temperature on the division of leaf protoplasts of *Lycopersicon esculentum* and *Lycopersicon peruvianum*. Plant Sci Lett 8:119–124

Zapata FJ, Sink KC, Cocking EC (1981) Callus formation from leaf mesophyll protoplasts of three *Lycopersicon* species: *L. esculentum* cv. Walter, *L. pimpinillifolium* and *L. hirsutum* f. *glabratum*. Plant Sci Lett 23:41–46

Addendum

Since this review was compiled, a large number of contributions have appeared which, at this late stage, cannot be collated in the main text. We present some salient points in the summary below.

Firstly, mention may be made of a few major monographs, e.g., by Giles (1983), Potrykus et al. (1983) and Vasil (1984). A monograph has also appeared on *Cereal Tissue and Cell Cultures*, highlighting problems in this group of plants, edited by Bright and Jones (1985). Protoplasts have continued to be choice material for investigating a number of physiological phenomena. For example, Gorton and Satter (1984) employed pulvinal protoplasts for analysis of movement of leaves and pinnae of *Samanea*, under the control of phytochrome and endogenous rhythms. Such problems as the mechanism of stomatal movements, or the types of ion involved or ion channel existing in the guard cells are also being approached through isolated protoplasts. In one contribution the "patch-clamping" technique has been applied to protoplasts (Moran et al. 1984).

However, closer to the theme of this article, several investigations have dealt with improvements in techniques of isolation and regeneration of protoplasts. That protoplasts can be separated on the basis of their surface charge by isoelectric focussing was shown by Griffing et al. (1985). Swanson et al. (1985) described a novel and efficient technique which employed a cyanogen bromide-activated Sepharose macrobead column, coupled with cellulase, for separating contaminating cells from protoplasts. Shneyour et al. (1984) have described another simple feeder layer technique for plating protoplasts at low density. In a further extension and refinement of their original microdrop technique, Koop and Schweiger (1985) have been able to regenerate whole plants from individually cultivated mesophyll protoplasts of *Nicotiana tabacum* in volumes as small as 20–80 nl. Negretiu et al. (1985) described techniques for the isolation of amino acid auxotrophs from protoplast cultures of *Nicotiana plumbaginifolia* and also contributed a critical review on the general problem of production and identification of biochemical mutants employing plant protoplasts (Negretiu et al. 1984).

Finally, a large number of new reports have appeared of organogenesis or of regeneration of embryoids or plantlets from protoplasts. As a few representative examples, mention must be made of *Brassica alba* (Glimelius 1984), *B. juncea* (Chatterjee et al. 1985), watermelon, i.e. *Cucumis sativus* (Orczyk and Malepszy 1985), *Tylophora indica* (Mhatre et al. 1984), *Broussonetia kazinoki* (Oka and Ohyama 1985), *Santalum album* (Rao and Ozias-Akins 1985), several species of *Solanum*, e.g. *S. aviculare* (Gleddie et al. 1985), *S. pennellii* (Hassanpour-Estahbanati and Demarly 1985) and *S. uporo* (Li and Constabel, 1984).

Among legumes, mention may be made of *Trifolium rubens* (Grosser and Collins 1985), *Medicago arborea* (Mariotti et al. 1984); *Psophocarpus tetragonolobus* (Wilson et al. 1985), *Glycine canescens* (Newell and Luu, 1985) and *Hedysarum coronarium* (Arcioni et al. 1985).

Several reports on monocots have also appeared. Toriyama and Hinata (1985) were able to obtain root formation and green spots (though no shoots) from protoplasts of anther-derived rice callus. Albino shoot regeneration was reported by Heyser (1984) in a millet, *Panicum miliaceum*. Additionally, cell division and callussing have been reported in *Sorghum bicolor* (Shourey and Sharpe, 1985), and oil palm, *Elaeis guineensis* (Bass and Hughes 1984), indicating that monocots are gradually becoming amenable to new protoplast isolation and cultural strategies.

A number of gymnosperms have also been studied by various investigators. However, success has been limited to colony formation or callussing.

References

Arcioni S, Mariotti D, Pezotti M (1985) *Hedysarum coronarium* L. In vitro conditions for plant regneration from protoplasts and callus of various explants. J Plant Physiol 121:141–148

Bass A, Hughes W (1984) Conditions for isolation and regeneration of viable protoplasts of oil palm (*Elaeis guineensis*). Plant Cell Rep 3:169–171

Bright SWJ, Jones MGK (eds) (1985) Cereal Tissue and Cell Culture. Nijhoff/Junk

Chatterjee G, Sikdar SR, Das S, Sen SK (1985) Regeneration of plantlets from mesophyll protoplasts of *Brassica juncea* (L) Czern. Plant Cell Rep 4:245–247
Chourey PS, Sharpe DZ (1985) Callus from protoplasts of *Sorghum* cell suspension cultures. Plant Sci 39:171–175
Giles KL (ed) (1983) Plant Protoplasts. Int Rev Cytol Suppl 16
Gleddie S, Keller WA, Setterfield G (1985) Plant regeneration from tissue, cell and protoplast cultures of several wild *Solanum* species. J Plant Physiol 119:405–418
Glimelius K (1984) High growth rate and regeneration capacity of hypocotyl protoplasts in some Brassicaceae. Physiol Plant 61:38–44
Gorton HL, Satter RL (1984) Extensor and flexor protoplasts from *Samanea* pulvini. II. X-ray analysis of potassium, chlorine, sulfur, phosphorus and calcium. Plant Physiol (Bethesda) 76:685–690
Griffing LR, Cutler AJ, Shargool PD, Fowke LC (1985) Isoelectric focussing of plant cell protoplasts. Plant Physiol (Bethesda) 77:765–769
Grosser JW, Collins GB (1985) Isolation and culture of *Trifolium rubens* protoplasts with whole plant regeneration. Plant Sci Lett 37:165–170
Hassanpour-Estahbanati A, Demarly Y (1985) Plant regeneration from protplasts of *Solanum pennellii*: Effect of photoperiod applied to donor plant. J Plant Physiol 121:171–174
Heyser JW (1984) Callus and shoot regeneration from protoplasts of Proso millet (*Panicum miliaceum* L.). Z Pflanzenphysiol 113:293–299
Koop H-U, Schweiger H-G (1985) Regeneration of plants from individually cultivated protoplasts using an improved microculture system. J Plant Physiol 121:245–257
Li GG, Constabel F (1984) Plant regeneration with callus and protoplasts of *Solanum uporo* Dun. J Plant Physiol 117:137–142
Mariotti D, Arcioni S, Pezzotti M (1984) Regeneration of *Medicago arborea* L. plants from tissue and protoplast cultures of different organ origin. Plant Sci Lett 37:149–156
Mhatre M, Bapat VA, Rao PS (1984) Plant regeneration in protoplast cultures of *Tylophora indica*. J Plant Physiol 115:231–235
Moran N, Ehrenstein G, Iwasa K, Bare C, Mischke C (1984) Ion channels in plasmalemma of wheat protoplasts. Science (Wash DC) 226:835–837
Negrutiu I, Brouwer D De, Dirks R, Jacobs M (1985) Amino acid auxotrophs from protoplasts cultures of *Nicotiana plumbaginifolia* Viviani I. BUdR enrichment selection, plant regeneration and general characterisation. Mol Gen Genet 199:330–337
Negrutiu I, Jacobs M, Caboche M (1984) Advances in somatic cell genetics of higher plants – the protoplast approach in basic studies on mutagenesis and isolation of biochemical mutants. Theor Appl Genet 67:289–304
Newell CA, Luu HT (1985) Protoplast culture and plant regeneration in *Glycine canescens* FJ Herm. Plant Cell Tissue Organ Culture 4:145–149
Oka S, Ohyama K (1985) Plant regeneration from leaf mesophyll protoplasts of *Broussonetia kazinoki* Sieb (Paper mulberry). J Plant Physiol 119:455–460
Orzyk W, Malepszy S (1985) In vitro culture of *Cucumis sativus* L. V. Stabilizing effect of glycine on leaf protplasts. Plant Cell Rep 4:269–273
Potrykus I, Harms CT, Hinnen A, Hutter R, King PJ, Shillito RD (eds) (1983) Proc 6th Int Protoplast Symp, Basel
Rao PS, Ozias-Akins P (1985) Plant regeneration through somatic embryogenesis in protoplast cultures of sandalwood, (*Santalum album* L.) Protoplasma 124:80–86
Shneyour Y, Zelcer A, Shamay I, Beckmann JA (1984) A simple feeder layer technique for the plating of plant cells and protoplasts at low density. Plant Sci Lett 33:293–302
Swanson EB, Wong RSC, Kemble RJ (1985) A novel method for the isolation and purificaton of protoplasts from friable, embryogenic corn (*Zea mays* L.) callus. Plant Sci 40:137–144
Toriyama K, Hinata K (1985) Cell suspension and protoplast culture in rice. Plant Sci 491:179–183
Vasil IK (ed) (1984) Cell culture and somatic cell genetics of plants Vol 1. Academic Press, London Toronto Tokyo
Wilson VM, Haq M, Evans PK (1985) Protoplast isolation, culture and plant regeneration in the winged bean, *Psophocarpus tetragonolobus* (L) DC. Plant Sci 41:61–68

II Protoplast Fusion and Early Development of Fusants

H. BINDING[1], G. KRUMBIEGEL-SCHROEREN[1], and R. NEHLS[2]

1 Introduction

The protoplasts of multicellular plants are normally enclosed in cell walls communicating by tiny plasmodesmata. Even in the absence of cell walls (e.g., in the pericarp of several solanaceae species), they sustain as individuals. The tendency of naked protoplasts to fuse is very low. This is mainly attributed to negative electric potentials of the surfaces (see Sect. 3.5.1). Nevertheless, fusion occurs in several systems in nature and can be induced, readily, in the laboratory.

This article is devoted to the various aspects of protoplast fusion in embryophytes. Whereas former reviews focused attention mainly on techniques (e.g., Vasil 1984) and on genetic aspects (e.g., Gleba and Sytnik 1984), the subject will be treated here preferentially from the viewpoint of physiology and developmental biology. After a brief survey of natural protoplast fusion, experimental protoplast fusion will be discussed in more detail.

2 Natural Plant Protoplast Fusion

Protoplast fusion in natural systems has been known for a long time, but so far, nearly no information is available on the mechanisms involved.

2.1 Fusion in the Sexual Cycle

Most commonly, cell fusion occurs as the initial step of zygote formation. The heterokaryotic state is established by fusion of either gametes or gametangiums. Two types of barriers must be overcome: at least one cell wall and plasma membranes. Little is known of the enzymatic functions that are needed for the removal of the cell wall material at the contact zones. The fusigenic conditioning of the plasmalemma is most likely established by changes of certain electric potentials of the surface and/or by alteration of the macromolecular structure of the membrane.

[1] Botanisches Institut der Christian-Albrechts-Universität, Biologiezentrum, Olshausenstraße 40–60, 2300 Kiel 1, FRG.

[2] PLANTA Angewandte Pflanzengenetik und Biotechnologie GmbH, Postfach 146, 3352 Einbeck, FRG

Results and Problems in Cell Differentiation 12
Differentiation of Protoplasts and of Transformed Plant Cells (Edited by J. Reinert and H. Binding)
© Springer-Verlag Berlin Heidelberg 1986

2.2 Fusion in Development

Protoplasts of higher plants are united to a symplastic continuum by plasmodesmata. They are supposed to be to a high degree residues of incomplete separation of daughter protoplasts during cytokinesis.

Formation of postgenital plasmodesmata has also been repeatedly described and/or discussed (for reviews see Jones 1976; Carr 1976). This process involves an event which can be interpreted as protoplast fusion of limited extent. Autoplastic connections are formed, for instance, during the formation of the false septum in the fruit of *Capsella bursa-pastoris* (Boeke 1973). Heteroplastic plasmodesmata occur in nature between plant parasites and hosts. This is indicated by electron micrographs, e.g., of haustoria of *Cuscuta* in host tissue (Kollmann and Dörr 1969; Fig. 1).

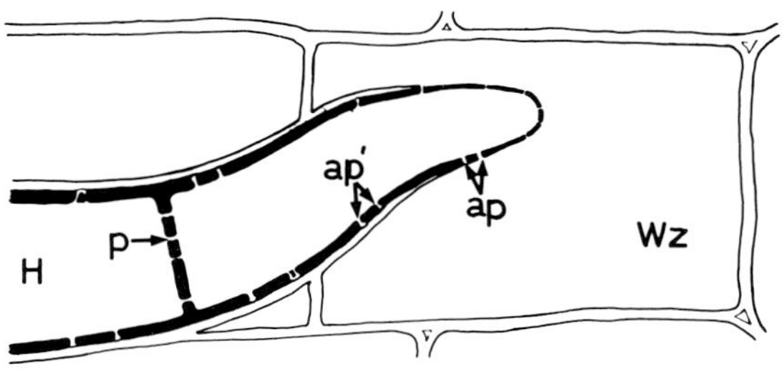

Fig. 1. Scheme of a *Cuscuta* hypha (*H*) in host cells (*Wz*), illustrating plasmodesmata between parasite and host (*ap*), closed by cell wall of the host formed later on (*ap*'). (Kollmann and Dörr 1969)

Neoformation of heterospecific plasmodesmata in an experimentally induced combination, namely, the periclinal chimera *Cytisus adami,* has been described previously by Buder in 1911 (cf. also Burgess 1972). Winkler (1938) discussed the establishment of "burdons" from grafting of tomato and black night shade by gene transfer through plasmodesmata. Unequivocal proof of plasmodesmata connecting stock and scion has been obtained recently by Kollmann and Glockmann (1985) by electron microscopic investigations of *Helianthus annuus* + *Vicia faba* grafts; the partner species exhibit significant properties for identification of the respective cells in the wound callus. Again, information on the mechanisms involved in the formation of postgenital plasmodesmata are completely lacking.

3 Experimental Protoplast Fusion

Protoplast fusion experiments are carried out to obtain information on the physical properties of the plasma membrane by investigations on the induction and the process of fusion; to study fusion bodies and the development of fusion products under physiological and genetic aspects; and to obtain plants with new genetic combinations of basic and applied interests.

It has been widely accepted in recent publications, to use the symbol " × " to differentiate somatic hybrids from sexual hybrids designated "(×)" and from graftings, "+".

3.1 Plasmolysis

Usually, plasmolysis is an essential prerequisite of experimental protoplast fusion as it is a part in the procedure to bring cells into fusable condition. Though it is routinely established in the experiments, insufficient attention is being payed to this process which bears some consequences on the possible composition of the fusants: It is well-known since the last century that the shapes of plasmolyzed protoplasts are different, depending on the cell types, the nature of the plasmolyticum, the osmolarity, and the duration of the incubation. Furthermore, it has been shown in the moss *Funaria hygrometrica* that it depends on the physiological state of the donor cells varied by particular culture conditions (Binding 1966). The various types of plasmolysis are easily visualized in gametophytes of archegoniates (Fig. 2). Convex plasmolysis normally leads to the formation of complete protoplasts (Fig. 2a). Hecht's filaments are very thin and frequently dissolved by time. Prolonged adhesion of protoplast in limited areas results in fractionation giving rise to small subprotoplasts besides a large one containing most of the cellular material (Fig. 2b). Sticking of the protoplasts over larger areas of the cell walls leads to convex plasmolytic shapes. The shrinking protoplasts either remain integer, contracting into one part of the cell (Fig. 2c) or – when two loci of positive plasmolysis are formed – divide into more or less equal portions of subprotoplasts of which only one contains the nucleus (Fig. 2d).

The consequences on the quality of protoplast preparations are evident: Integer protoplasts are reliably obtained from cells plasmolyzed as demonstrated in Fig. 2a (see Sect. 3.4.2); sticking to the cell walls (Fig. 2c, d) facilitates the fusion

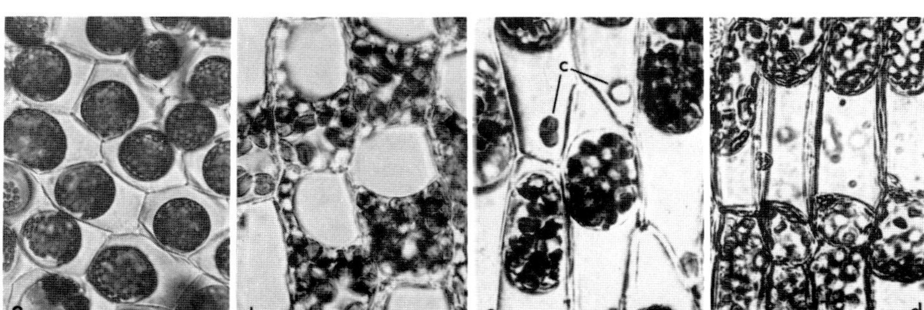

Fig. 2 a–d. Shapes of plasmolysis in seawater diluted to approx. 70%. **a** Perfect convex plasmolysis in a prothallium of the fern *Woodsia pulchella*; **b** concave plasmolysis in a leaflet of *Funaria hygrometrica*; **c** convex plasmolysis and cytoplasts (*c*) in a leaflet of *Physcomitrium piriforme*; **d** subprotoplasts in a leaflet of *Bryum erythrocarpum* (Binding, unpubl.)

of neighboring protoplasts during cell wall digestion (see Sect. 3.3) and may be responsible for reduced yields of viable protoplasts; cells with fractionated protoplasts (Fig. 2b, d) are useful for the isolation of subprotoplasts (see Sect. 3.4.4).

3.2 Fusion Within Cell Walls

In some early investigations, subprotoplasts have been fused which were created by plasmolysis and left inside the entire cell walls. However, in these cases it can never be excluded that Hecht's filaments were conserved and, hence, the subprotoplasts were never completely separated. Fusion of protoplasts of two unilaterally opened cells which has been immensely successful, for instance in *Acetabularia* (Hämmerling 1963) and *Phycomyces* (see Ootaki et al. 1977) has never been observed in higher plants.

3.3 Spontaneous Fusion

In most of the cases, suspensions of isolated protoplasts contain multinuclear bodies. It is strongly suggested that they are formed by fusion of neighboring cells during enzymatic digestion. The process has been termed "spontaneous fusion" (Withers and Cocking 1972). A fusion body is shown in Fig. 3.

The frequencies vary from species to species and are dependent on the donor tissues as well as the isolation conditions. They have never been found after mechanical isolation of moss protoplasts during the investigations described in 1966 (Binding 1966) nor in *Sphaerocarpos* after enzymatic isolation (Schieder 1974). High degrees of multinuclear protoplasts are commonly found in preparations of species of the Magnoliatae. As much as 30% of the protoplasts carried more than

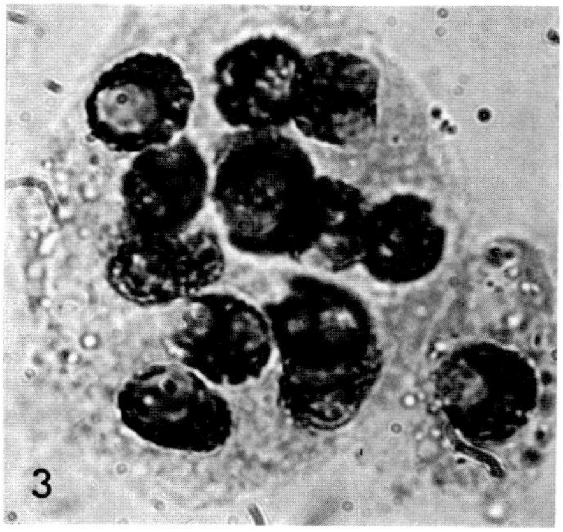

Fig. 3. Spontaneous fusion. Fusion body of *Vicia faba* shoot apex, plasmolyzed by 0.5 M mannitol during enzyme incubation (unpubl.)

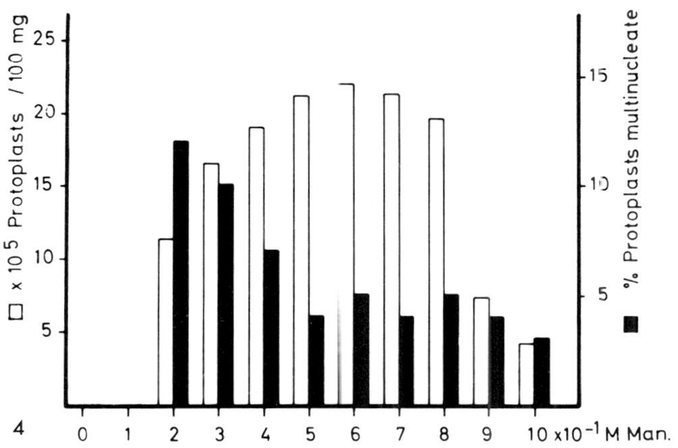

4

Fig. 4. Spontaneous fusion. Degrees of multinucleate mesophyll protoplasts of *Petunia hybrida;* preincubation in different mannitol solutions; enzymatic digestion of cell walls in the presence of 0.6 M mannitol (Binding 1974 b).

one nucleus in tobacco (Power and Frearson 1973); up to 12% were reported for *Petunia hybrida* (Binding 1974 b); and an average of 8% was observed in *Datura innoxia* (Schieder 1976).

The multinuclear bodies are most likely formed by the expansion of plasmodesmata which have not been broken by plasmolysis (Withers and Cocking 1972). Strong plasmolysis under conditions favoring convex plasmolysis is, therefore, supposed to reduce the degree of multinuclear protoplasts. The effect of different osmolarities of the plasmolyticum prior to enzyme incubation is demonstrated by Fig. 4 (Binding 1974 a). The correlation of convex plasmolysis and absence of multinucleate protoplasts is given in *Sphaerocarpos* and may be deduced from the appearance of a plasmolyzed fern prothallium (Fig. 2 a).

A rather different type of spontaneous fusion was reported for protoplasts from meiotic pollen mother cells of two liliaceous species (Ito and Maeda 1973). These protoplasts fused after release from the donor tissue just upon physical contact without being externally influenced. The ability to form fusion bodies was restricted to a short period after isolation and the frequencies were dependent on the meiotic stages.

Boss et al. (1983) reported on protoplasts of carrot conditioned to fuse by their developmental stage. These carried thread-like protrusions and exhibited altered membrane fluidity as monitored by electron spin resonance.

Recently, fusants were observed after culture of mixed protoplast suspensions of a green and a plastid mutant albino strain of *Solanum nigrum* at high density in agarose media (Binding 1984; Binding and Kollmann 1985). The fusant nature was indicated by characteristic mosaic patterns which were found not only in the original fusant, but also in adventitious shoots and in seedlings (Binding and Kollmann, unpubl.). The time and process of fusion in coculture are being investigated.

In rare cases, the presence of binuclear protoplasts must not be explained solely by spontaneous fusion. Wallin et al. (1974) were able to attribute the phenomenon, at least partly, to mitotic activity in the course of protoplast isolation. There was always a strict dependence on the mitotic frequencies in the cell suspensions used as donor material.

Spontaneous fusion is generally regarded as being of no practical applicability with respect to basic genetic research or breeding (Schieder and Vasil 1980). Moreover, it is even an experimental drawback when low chromosome numbers in the fusion products are intended. But there might be, admittedly, a chance for heterospecific fusion by spontaneous fusion during the protoplast isolation from periclinal chimeras formed by grafting somehow renewing the aspect of "burdons" (Winkler 1938; Brabec 1954). An important prerequisite would be the occurrence of spontaneous fusion between protoplasts of two layers of the apex. However, the chance seems to be rather low as shown from investigations with chlorophyll deficient mutants of chimeric nature of *Solanum nigrum* and *Petunia hybrida* (Binding et al. 1982b). No mosaic was obtained in more than 1000 individual protoplast clones.

3.4 Fusion of Isolated Protoplasts

Protoplast fusion as a tool for experimental hybridization has already been proposed by Küster (1910). Unfortunately, his experiments suffered from the inappropriate cell culture techniques of his days; methods for the preparation of protoplasts (Klercker 1892) were already available, and he himself had detected agents for protoplast fusion.

3.4.1 Plant Protoplasts Suited for Fusion

Phenomena concerning the isolation of viable plant protoplasts are surveyed in the preceding article. Technical details have been discussed in several publications (e.g., Binding and Nehls 1982; Vasil 1984). As the success of protoplast fusion experiments mostly depends on the properties of the protoplast preparations, some particularly interesting features with regard to protoplast fusion will be considered here.

In general, all structures that are surrounded by a plasma membrane – or even liposomes – and that are stable enough to tolerate the fusion procedure, are especially suited since highly efficient fusion techniques are available. It must be realized when the types of protoplasts for a fusion experiment are chosen that the ability to regenerate varies largely even within a given species (e.g., Binding et al. 1978).

Additional effects were obtained in somatic cell hybridization experiments. In several systems (e.g., Constabel et al. 1975a) cell colonies were formed even if only one of the parental protoplast types was able to regenerate under the experimental conditions. Maliga et al. (1977) found restoration of organogenic potencies in fusion products of strains of *Nicotiana sylvestris* and *N. knightiana* which both were incapable of shoot formation. However, it seems to be advisable not to rely upon new regenerative capacities in fusion products but, if possible, to use at least one protoplast type which is easily regenerated.

Particular high stability of the isolated protoplasts is needed for fusion experiments. The stabilities mainly depend on the physiological condition of the original cells. This was initially found in mosses (Binding 1966) and has been confirmed in numerous investigations in higher plants. The establishment of well-adapted cells or tissues is managed by controlled environments and by using certain types of tissues or cells.

3.4.2 Protoplast Preparation

The protoplasts used in early fusion experiments have been isolated by mechanical methods (e.g., Küster 1910; Hofmeister 1954; Binding 1966). In the cases of fruit endosperms (in tomato, black nightshade, and others), they are already free of cell walls and are purified prior to use in fusion experiments (Binding 1976). Since 1970 (Power et al.), protoplasts for somatic hybridization have been isolated nearly exclusively by enzymatic digestion of the cell walls as introduced by Cocking (1960).

The preparations should be essentially free of debris and the protoplasts must be in good condition in order to resist the fusion procedure. The time between the removal of the hydrolytic enzymes and the protoplast fusion must be kept as short as possible to ensure that the protoplast surface is still perfectly free of cellulose fibers. For instance, Weber et al. (1976) found a drastic decrease in the fusion efficiencies as early as 15 min after the removal of the cell wall degrading enzymes; after 2 h, the fusion rate was reduced from 9% (after 5 min) to only 1%. The sensibility of Calcofluor White staining which was often used to prove the protoplast state was far below the minute amounts of freshly synthesized cellulose capable of inhibiting membrane fusion. Calcofluor-detectable depositions did not appear earlier than after 4 h of culture. However, according to ultrastructural studies with *Vicia hajastana,* it was found that the formation of a fine network of microfibers had occurred on the membranes of protoplasts which were fixed 10 min after they had been removed from the enzyme solution (Williamson et al. 1977).

3.4.3 Characteristics of Protoplasts Combined in Fusion Experiments

3.4.3.1 Genetic Peculiarities

The combination of two genetically different protoplasts has a number of implications: the choice of protoplast types is determined in several cases, especially in plant breeding, by the intention of achieving a specific new constellation of genetic information; particular combinations are also devised or, at least, utilized for diverse demands in the course of somatic hybridization experiments, such as recognition and clearcut identification, selection, and the analysis of the nuclear and the plasmonic constitution of the fusants in different stages of development. Markers suited for one or more of these purposes are, for instance, structure

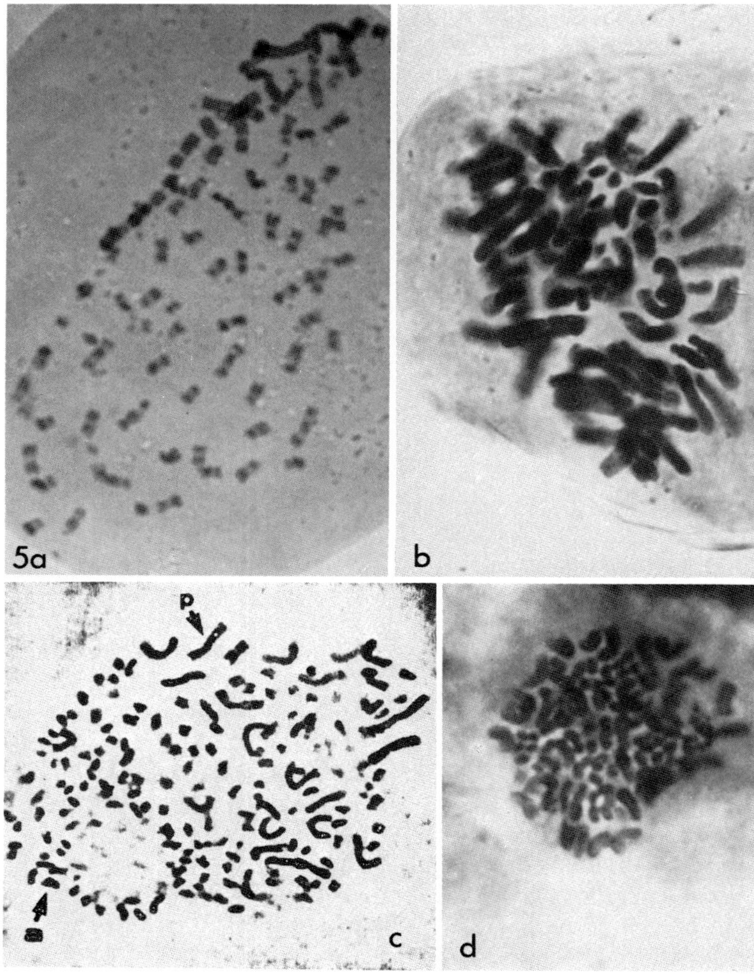

Fig. 5 a–d. Mitoses in fusion products illustrating the use of chromosome sizes (large/small) for investigations in somatic hybridization. **a** *Glycine max* – small – (×) *Nicotiana glauca* – large – (Kao 1977); **b** *Vicia faba* – large – (×) *Petunia hybrida* – small – (Binding and Nehls 1978); **c** *Atropa belladonna* – small – (×) *Petunia hybrida* – large – (Gosch and Reinert 1978); **d** *Atropa belladonna* – small – (×) *Datura innoxia* – large – (Krumbiegel and Schieder 1979)

(Figs. 13e, 20) and pigmentation of plastids, organization of the interphase nuclei (Figs. 13c, 13d, 13e, 17, 18), shape of the chromosomes (Fig. 5), density of the cytoplasm (Fig. 7), auxotrophies, resistances, developmental potencies, protein patterns, restriction fragment patterns of DNA, and morphological peculiarities of the fusants. Some details on the application of marker systems in early development will be given in Section 4.2. Their benefits for selection and analysis in more advanced stages will be presented in Part III of this volume.

3.4.3.2 *Characteristics of Cell Differentiation*

Differential characteristics based on the developmental stages of the donor cells are useful for investigations on the formation and early development as well as for the purpose of mechanical selection of fusion bodies (Fig. 14). In a number of investigations mesophyll protoplasts containing well-differentiated chloroplasts have been paired with colorless protoplasts of cell suspensions (Figs. 8, 11, 13a, 13d, 14), callus (Fig. 12) or apices (Figs. 9, 10, 13b, 20), or with protoplasts with pigmented vacuoles (Potrykus 1971) or plastids (Binding 1976). Differences in buoyant densities are appropriate for the separation of fusion bodies in density gradients (Fig. 15). It has been pointed out earlier that pairs of developmental markers are easily established in combinations of nearly any genetic constitutions and, hence, are highly useful in plant breeding programs (e.g., Binding and Nehls 1978).

3.4.3.3 *Artificially Established Specific Protoplast Properties*

Vital staining of protoplasts by fluorescent dyes (Galbraith and Galbraith 1979; Galbraith and Mauch 1980; Galbraith and Harkins 1982; Patnaik et al. 1982; Berry 1983; Galbrait et al. 1984; Harkins and Galbraith 1984; methods: Galbraith 1984) is universally applicable for the detection of fusion bodies and also for automatic selection. The parental protoplasts were labeled by different fluorescent dyes or only one type was labeled artificially and chlorophyll fluorescence was used as the counterpart. Staining did not affect the viability of the protoplasts.

Metabolic inhibition of parental cells and subsequent complementation in fused cells which was developed for animal cells by Wright (1978) was introduced to plant systems for early selection by Nehls (1978). In particular, a system of iodoacetate and diethylpyrocarbonate was used (Fig. 18; Nehls 1978, 1981; Nehls and Binding 1979). Unilateral blocking by iodoacetate has been applied by Nehls and Binding (1979), Nehls (1981), Sidorov et al. (1981), Wallin and Savage (1982), and Cella et al. (1983).

Reversion of the ζ-potential to positive charges of the plasmalemma of one of the protoplast types has been devised to obtain controlled heteroplastic fusion (Nagata et al. 1979; cf. Sect. 3.5.1). An additional possibility of modifying protoplasts is provided by X-irradiation which causes functional elimination of the nuclei (cf. Sect. 3.4.4.1).

3.4.4 Subprotoplasts

The fusion of two protoplasts combines the complete nuclear and cytoplasmic genetic information of two cells. The genetic complexity of fusion bodies is reduced during subsequent development: plastids and mitochondria segregate; the nuclei may be unequally distributed to the daughter cells, for instance, when only one nucleus performs the mitotic cycle, while the other one is resting; and, finally, chromosomes may be lost in mitoses of fusant clones (see Sects. 4.2.2, 4.2.3). Consequently, various types of cell lines can be isolated.

However, reasons exist for trying to establish certain organelle combinations just by the fusion experiment, i.e., that protoplasts lacking special types of organelles are used for fusion. Such experiments may be indicated when a desired recombinant is not characterized by easily detectable genetic markers or when the fusion body already should be free of certain organelles which would interfere with the development of the fusion products or with the origination or proliferation of just the desired cell line. A lower complexity of the fusion body is achieved by using protoplasmic structures which contain only parts of the genetic information. This condition is met by inactivation of certain cell organelles (e.g., the nucleus by X-rays) and by subprotoplasts. Some of these may play the role of "carriers of selected types of organelles" (Binding and Kollmann 1976; see also Binding 1976, 1979, Lörz and Potrykus 1980). Subprotoplasts consisting of a nucleus surrounded by a small portion of cytoplasm were termed "miniprotoplasts" (Wallin et al. 1979), those representing nucleus-free cell fragments "cytoplasts" (Lörz et al. 1981).

3.4.4.1 Formation of Isolated Subprotoplasts

Subprotoplasts with and without nuclei have been described in several reports. They are formed naturally, for instance, in the juicy pericarp of several plant species (e.g., tomato; *Solanum nigrum*). Their formation is shown schematically in Fig. 6a (Binding and Kollmann 1976).

Fractionation of protoplasts also occurs during plasmolysis (see Sect. 3.1). As a consequence, subprotoplasts are found in almost any protoplast preparation (Binding and Nehls 1979). Particularly high degrees are produced in prosenchymatous cells which have been used in experiments with mosses (Binding 1966) and some species of the solanaceae family (Binding and Nehls 1980). The protoplasts were isolated in both cases by compression of dissected tissues. Lörz and Potrykus (1980) described the formation of miniprotoplasts and cytoplasts when growing pollen tubes of *Hyoscyamus muticus* and *Nicotiana tabacum* were treated with wall-degrading enzymes. Vatsya and Bhaskaran (1981) emphasized the correlation between the formation of cytoplasts and osmolarities of the enzyme solution used for protoplast isolation from cotyledonary leaves of *Brassica oleracea;* the percentages of subprotoplast formation increased with the hypertony of the enzyme solutions.

Wallin et al. (1978) used a technique with cytochalasin B known from animal cells for enucleation of isolated protoplasts. In combination with high speed centrifugation they were able to produce miniprotoplasts and cytoplasts in carrot, tobacco, and pea. In a slightly modified method, this technique was applied in the preparation of enucleated protoplasts and miniprotoplasts from *Allium cepa* (Bracha and Sher 1981). The formation of cytoplasts and miniprotoplasts from cultured cells of *Zea mays, Hyoscyamus muticus,* and *Nicotiana tabacum* was achieved using a discontinuous isoosmotic density gradient (Lörz et al. 1981). The protoplasts and – later – the cytoplasts were retained at a phase boundary, whereas the nuclei enclosed in a thin cytoplasmic layer were sedimented as miniprotoplasts to the bottom of the centrifuge tube.

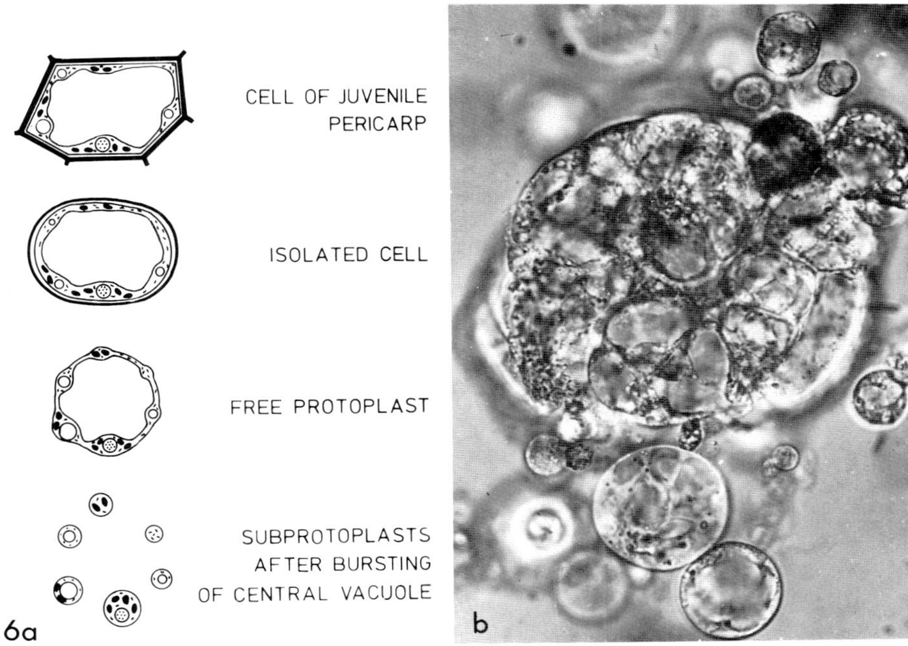

Fig. 6 a, b. Subprotoplast formation. **a** Subprotoplast formation of a pericarp cell of tomato (Binding and Kollmann 1976); **b** subprotoplasts formed by budding of a protoplast of *Kentranthus ruber,* 6 d in culture (unpubl.)

It has been frequently observed that protrusions (buddings) are formed by protoplasts in culture which, after separation, give rise to subprotoplasts. In a number of cases, budding is easily explained by the generation of an osmotic gradient from the protoplasts to the culture media during regeneration of a cell wall (Binding 1966). In other cases, ist seems to be a more active excretion (Lörz and Potrykus 1980) resembling the process of physiological plasmoptysis which was observed in pollen tubes and several other types of cells (Küster 1958). This gave rise either to various types of cytoplasts (Fig. 6 b) or to miniprotoplasts (Binding and Kollmann 1976). The mechanism of this type of budding is completely unknown.

Mechanical fractionation of protoplasts occurs when agglutinated protoplasts (e.g., in a solution of polyethylene glycol: see Sect. 3.5.2) are driven apart by a flow (Fig. 9; Binding and Nehls 1980). It is interesting to know that a high probability of protoplast-to-subprotoplast fusion is given in nearly any protoplast incubation for fusion as a consequence of parting of protoplasts during plasmolysis and of mechanical fractionation (see Sect. 4.2.2.2).

In order to achieve functional enucleation, Zelcer et al. (1978), Aviv and Galun (1980), Galun et al. (1982), and Gupta et al. (1982, 1984) irradiated protoplasts which resulted in an inactivation of the nuclei with higher probability than other cell organelles.

3.4.4.2 Enrichment and Separation of Subprotoplasts

Step gradient centrifugation has been used for the collection of subprotoplasts and for the separation of different types of subprotoplasts from entire protoplasts and from one another. Enrichment of pericarp cytoplasts was obtained by using sucrose solution and culture medium (Binding 1976). A more sophisticated step gradient was constructed by the use of percoll, mannitol, and $CaCl_2$ for subprotoplasts of cell suspension cultures (Lörz et al. 1981; Archer et al. 1982; see also Sect. 4.1 and Fig. 15).

3.4.4.3 Metabolic Activities of Subprotoplasts

Already in 1897, Townsend investigated cell wall formation in plasmolyzed cells. He found that subprotoplasts were able to regenerate cell walls either when they contained the nucleus or when they were connected by cytoplasmic strands to a miniprotoplast or to a protoplast in the neighbour cell. Accordingly, isolated nucleus-free cytoplasts never formed cell walls (Binding 1966; Lörz et al. 1981; Archer et al. 1982). On the other hand, miniprotoplasts were able to regenerate: the miniprotoplasts prepared by treatment with cytochalasin B increased in volume during 24 h of culture up to the size of protoplasts and divided (Wallin et al. 1978). The metabolic activity of nuclear subprotoplasts enriched by centrifugation in a density gradient with percoll proved to be very high; this was indicated by measurements of protein synthesis (Lörz et al. 1981). They also increased rapidly in size within a short cultivation time and, with delay, formed cell colonies.

3.5 Processes Involved in Protoplast Fusion

Protoplast fusion can be induced by highly efficient procedures. Once pure protoplast suspensions are prepared, there seem to be really no limitations to the formation of fusion bodies from protoplasts of any origin in angiosperms (Constabel et al. 1975b) and even between higher plant protoplasts and animal cells (e.g., Jones et al. 1976; Willis et al. 1977). No barriers of incompatibilities have been observed affecting the fusion process.

Successful combination of even far remote systems have contributed to the concept of a common basis of the physical processes of membrane fusion of living cells. A number of details are still poorly understood. However, the information and considerations are sufficient to encourage the elaboration of more sophisticated methods of protoplast fusion as this is already supported by the development of a procedure for directed heterologous fusion by reversion of the zeta-potentials of the protoplasts of one of the parents (Nagata et al. 1979).

3.5.1 Physiological Mechanisms of Induced Protoplast Agglutination and Fusion

Fusion of isolated plant protoplasts is possible only under particular physiological conditions. The intrinsic fusion process is preceded by the establishment of tight contact between the protoplasts.

The association of isolated protoplasts has been described in early literature as "fusion like soap bubbles" (Hofmeister 1954), as "plasmosyndesis" (Binding 1966), and is presently termed "agglutination". It can be induced by a number of reagents, for instance, by seawater (Fig. 7; Binding 1966), calcium ions (Keller and Melchers 1973), polyethylene glycol (PEG; Figs. 8, 9, 10, 11; Kao and Michayluk 1974), and a number of other polymers of neutral or polycationic nature.

Agglutination is mainly attributed to the removal of negative surface charges of isolated protoplasts which have been claimed by Ruesink (1971). Nagata and Melchers (1978) confirmed the occurrence of these charges, termed ζ-potentials, by means of electrophoresis. ζ-Potentials were obtained which ranged between -10 mV to -35 mV, depending on the species and the ploidy level, but independent of age, season, and culture conditions. The authors were able to remove the charges by acidic phosphatase treatment which provides strong evidence for the participation of membrane-bound phosphate groups in generating the negative potentials. The correlation between induction of agglutination and removal of the negative ζ-potentials is clearly indicated by an observation of Nagata and Melchers (1978). They found that calcium ions extinguished any net charge at a concentration of 100 mM which is sufficient for ready agglutination and fusion.

The removal of the negative potentials accounts for the abolishment of repulsion of isolated protoplasts from one another. However, this would not suffice to lead to the tight contact between agglutinated protoplasts. The respective forces are most likely the same as for animal cell adhesion. Curtis (1960) considered the importance of the action of van der Waal's forces which are normally masked by the negative surface charges (cf. also Poste and Allison 1971, 1973). Nagata et al. (1979) created additional attractive forces by the incorporation of positively charged phospholipids into the plasmalemmas of one of the fusion partners.

Polyethylene glycol (PEG) is an efficient agglutination-inducing agent. It is a nonionic weak surfactant. Protoplast fusion occurs mainly after its dilution in the incubation mixture (Burgess and Fleming 1974). The action of PEG is not yet sufficiently understood. PEG in concentrations which are necessary for agglutination is more or less harmful to the protoplasts. The incubation must be, therefore, restricted, often to just a few minutes. On this basis, it can be even used as a selective blocker impeding the regeneration of one of the parental protoplast types (e.g., in potato; Binding et al. 1982a).

Agglutination is also obtained by other substances, possibly with less negative side effects. Dextranesulphate seems to be particularly well suited (Kameya 1979, 1982; Senda et al. 1982).

Establishing conditions for agglutination does not suffice for efficient protoplast fusion. This can be, for instance, concluded from investigations on the combination of protoplasts carrying opposite polarities (Nagata et al. 1979). High fusion rates were not obtained before the addition of calcium ions.

The action of calcium is not restricted to the removal of the negative potentials. Boss and Mott (1980) discovered a considerable increase in membrane fluidity. The effect could be suppressed by the addition of chelating agents. A correlation of cell fusion and alterations of the membrane fluidity had already been

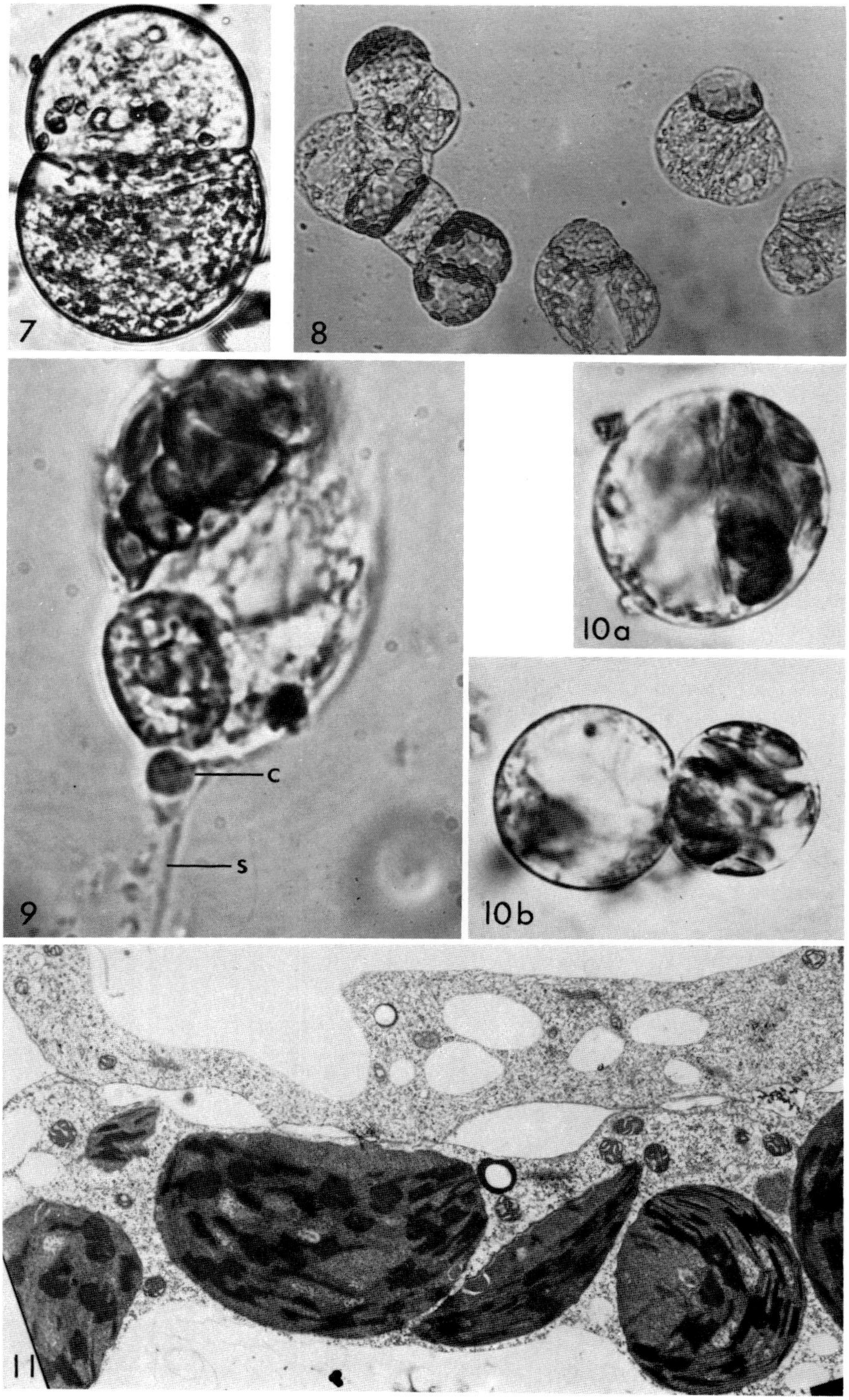
7
8
c
s
9
10a
10b
11

considered in animal cell fusion (hen erythrocytes; Ahkong et al. 1975). The authors offered a hypothetical model for cell fusion: Upon the influence of an exogeneous chemical agent, the lipid bilayer of the membrane is perturbed and this is associated with an increase of its fluidity. In the extreme case, lipid micelles might be formed, and proteins and glycoproteins are excluded from that region. Finally, the perturbed lipid layers of adjacent membranes together with other constituents rearrange to proper membrane architecture leaving small cytoplasmic bridges.

Temperature has a significant influence on the fusion rate. Wallin et al. (1974) observed a better protoplast aggregation at 15 °C rather than at 25 °C or 35 °C. Fusion, however, was most efficient at the higher temperatures. Similar results were obtained by Burgess and Fleming (1974) and Senda et al. (1980). The latter observed an increasing effect on membrane fluidity which is again in accordance with the theories mentioned above.

The application of high pH values to fusion in combination with calcium ions is highly efficient. This was discovered by Keller and Melchers (1973) and is widely and successfully used. Most likely, it plays an additive role in the alteration of the electric surface loads, but real knowledge especially on its action in fusion is not available.

The influence of electric fields on protoplast fusion has been investigated in several laboratories. The application of a low direct current to seawater resulted in fusion of moss protoplasts (Binding 1966). In this case, it may be discussed if the current itself, or alterations of pH or ion distribution as a consequence of electrolysis of the seawater were responsible Zimmermann and Scheurich (1981) introduced a technique for protoplast fusion which is based on the action of alternating electric fields. The principle of this technique includes a reversible electrical breakdown of the membranes upon polarization to about 1 V within nano- to milliseconds. The alternation of the field at high frequencies masks the negative surface potentials. Furthermore, the membrane conductivity and permeability is increased extensively. As the applied electric field is not uniform, the affected protoplasts, therefore, behave like dipoles which tend to move in the direction of the increasing field. In the course of their dielectrophoretic movement, the protoplasts attract each other according to their dipole characteristics, form pearl

Fig. 7. Protoplast agglutination. *Funaria hygrometrica* + *Bryum erythrocarpum* – dense – in seawater (Binding 1966)

Fig. 8. Protoplast agglutination. *Pisum sativum* – mesophyll – + *Vicia hajastana* – cell suspension – in PEG (Kao and Michayluk 1974)

Fig. 9. Protoplast agglutination. *Vicia faba* – apex – + *Petunia hybrida* – mesophyll – in PEG; tearing off one *Petunia* protoplast (out of the field of view) caused formation of a cytoplasmic strand *s* and of a cytoplast containing one *Petunia* chloroplast *c* attached to the *Vicia* protoplast (Binding and Nehls 1980)

Fig. 10 a, b. Protoplast agglutination. *Vicia faba* – apex – + *Petunia hybrida* – mesophyll –; **a** in PEG; **b** after dilution of PEG (Binding and Nehls, unpubl.)

Fig. 11. Protoplast agglutination. *Pisum sativum* – mesophyll – + *Vicia hajastana* – cell suspension –; electron micrograph of the contact zone (Fowke and Gamborg 1980)

chain structures, and fuse mainly to giant fusion bodies. This technique has been modified by additionally using calcium and several polymers causing agglutination (Senda et al. 1982). Koop (Koop et al. 1983; Koop 1984) refined the fusion technique by the use of electric fields so that fusion of single pairs of selected protoplasts and regeneration of fusant plants occured.

3.5.2 Morphological Processes in Agglutination and Fusion

Normally, intrinsic protoplast fusion is preceded by agglutination (Figs. 7, 8). Agglutinated protoplasts adhere to one another so tightly that withdrawal causes the formation of cytoplasmic threads between them and even the fractionation of one of the partners (Fig. 9). Despite this tight association, most of the agglutinated protoplasts dissociate with time, especially upon changes of the osmotic situation (Binding 1966) or removal of the agglutinating agent. This is even true when the associated protoplasts appear as an entire round unit (Fig. 10). Figure 11 shows an electron micrograph demonstrating that the plasma membranes initially adhere only across small areas which are separated by larger areas of lens-shaped or flat gaps (Burgess and Fleming 1974; Fowke et al. 1975a, 1977).

Protoplast fusion is a fast process only in particular cases, for instance, when glass splinters are used (Binding 1966) or when electric fields are applied. Usually, it takes long periods of time, probably up to 1 h or more. This is indicated by the gradual increase of the numbers of fusion bodies during incubation in fusion-inducing agents (Fig. 12) as well as after removal of these agents. Electron micrographs indicate that the initially tiny contact zones between agglutinated protoplasts enlarge and then narrow fusion channels appear (Burgess and Fleming

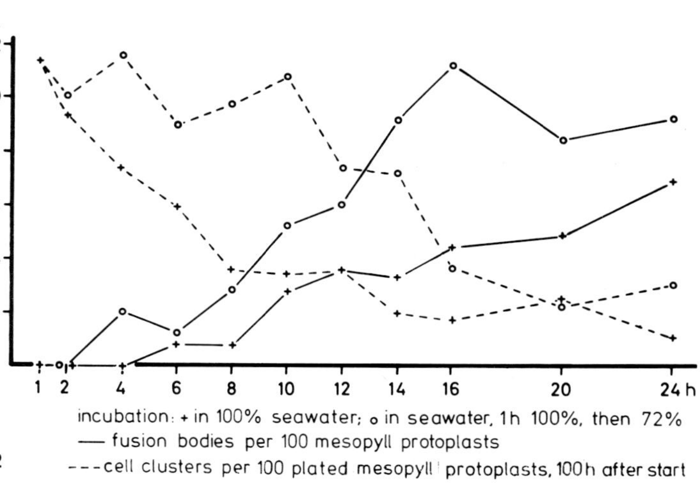

12

Fig. 12. Fusion of *Petunia hybrida* – mesophyll – (×) *P. hybrida* S2 – streptomycin resistant cell culture – in seawater pH 6.0; the weak fusigenic activity at low pH favored the visualization of the gradual increase of the numbers of fusion bodies correlated to a decrease in survival (Binding 1974a).

1974; Fowke et al. 1975b). Finally, the cytoplasmic connections expand to give rise to a true fusion body. It is supposed that the included parts of the plasmalemmas form vesicles which disperse in the cytoplasm (Fowke et al. 1975a).

4 Fusants

4.1 Peculiarities of Fusants for Recognition and Selection

The protoplast populations resulting from a fusion experiment are composed of parental protoplasts derived from single cells, spontaneous fusion bodies, and induced fusants, if no single pair of protoplasts has been treated in a microdroplet (Koop et al. 1983). Hence, usually differential recognition of the fusants is an essential prerequisite for investigations on their development. Various markers and growth characteristics have been utilized for this purpose (see Sect. 3.4.3). Investigations in early stages up to small cell clusters have been possible by differential visual markers of the parental protoplasts transmitted to the fusants (Fig. 13), by relative survival of the fusants (see Sect. 3.4.3), and also by fusant-specific development of cell organelles (see Sect. 4.2.2.). The transmitted visual markers enabled additionally the selection of microdroplets with individual fusants (Kao 1977) and the separation of fusants from mixed protoplast populations by micropipettes (Fig. 14; Gleba and Hoffmann 1978; Menczel et al. 1978; Nehls 1981; Sidorov et al. 1981; Gleba et al. 1982; Hein et al. 1982; Patnaik et al. 1982) and by a cell sorter (Galbraith and Harkins 1982; Galbraith et al. 1984; Harkins and Galbraith 1984). Buoyant densities of fusion bodies and uniparental protoplasts have been exploited by Harms (1977) and Harms and Potrykus (1978) for the separation in step gradient centrifugation (Fig. 15).

4.2 Early Development of Fusants

Fusion combinations established so far are compiled in Table 1. Early developmental processes have been investigated only in a relatively small number of cases. The behavior of nuclei and plastids indicate the initiation of development finally leading to the clonal variation of fusion products which is treated in Part III of this volume. Some of the processes are illustrated in Fig. 16. The behavior of mitochondria in young fusants has not been examined because of the lack of appropriate markers; the available analytical methods allow investigations on mitochondrial traits in fusant clones only when they are grown to larger biomass (e.g., DNA restrictase patterns) or to flowering plants (male sterility).

4.2.1 Cell Wall Formation

Protoplasts with regenerated walls (plastocytes; Binding 1966) have been obtained in most of the investigated combinations (cf. Table 1). Several factors, depending on the respective experiment, may have been responsible for failure of

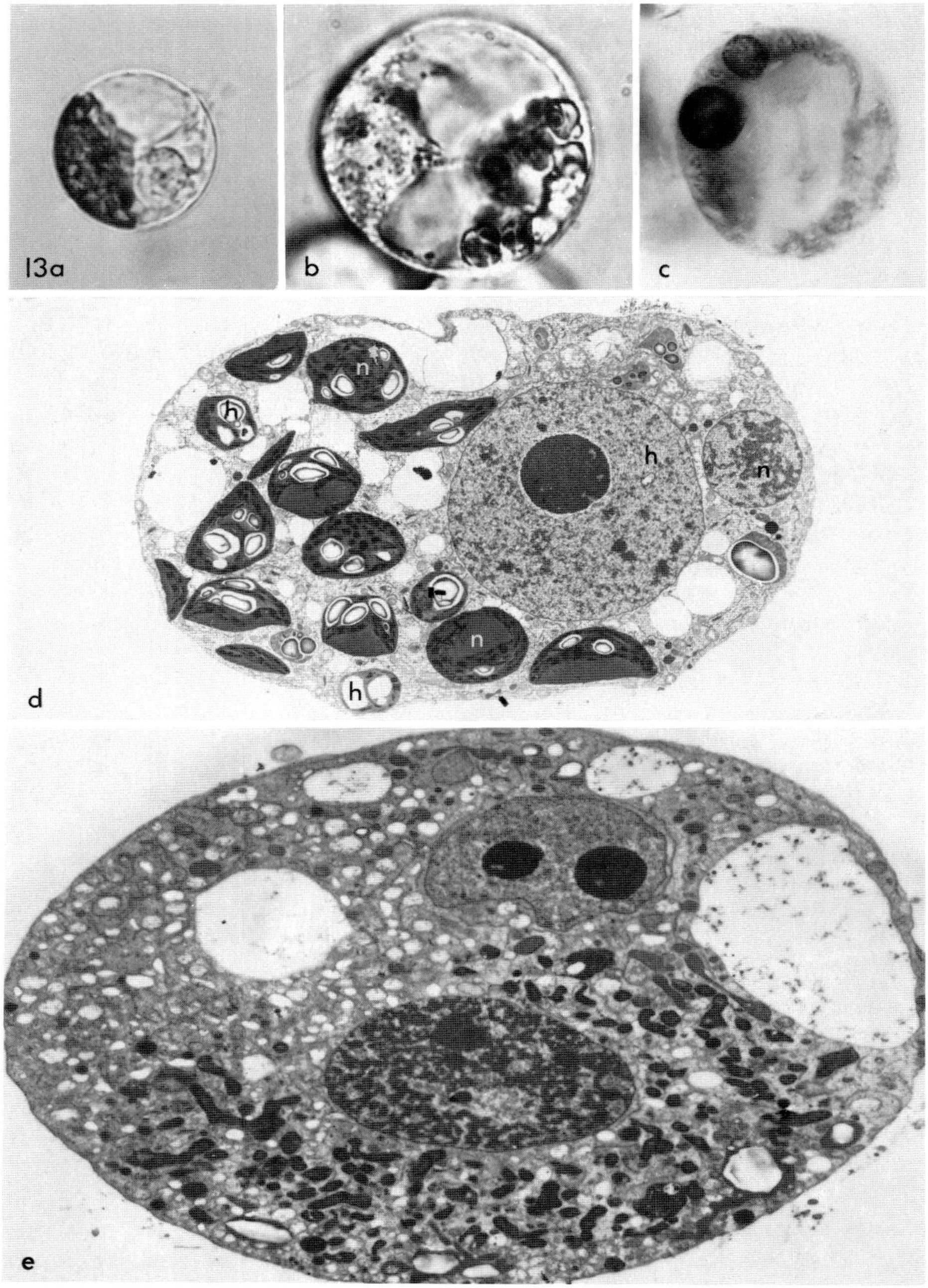

Fig. 13 a–e. Fusion bodies. **a** *Hordeum vulgare* – mesophyll – (×) *Glycine max* – cell suspension – (Kao and Michayluk 1974); **b** *Vicia faba* – apex – (×) *Petunia hybrida* – mesophyll – (Binding and Nehls 1978); **c** same as **b**, stained by acetocarmine (Binding and Nehls, unpubl.); **d** *Vicia narbonensis* – mesophyll, *n* – (×) *Vicia hajastana* – cell suspension, *h* – (Rennie et al. 1980); **e** *Vicia faba* – root nodule – (×) *Daucus carota* – cell suspension – (Davey et al. 1980)

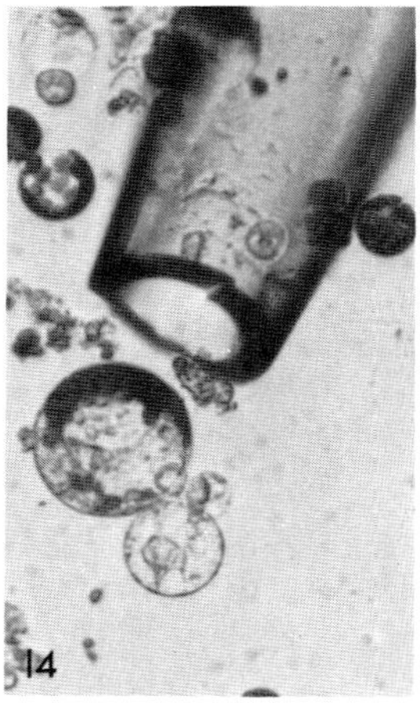

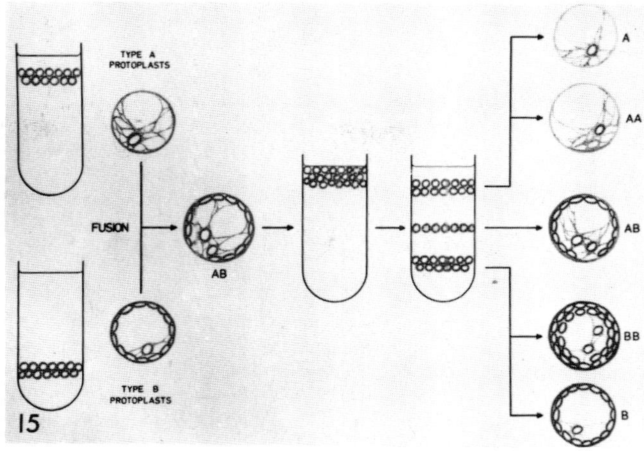

Fig. 14. Separation of fusion bodies from the bulk of protoplasts. Manual separation by a pipette; *Nicotiana tabacum* – cell suspension – (×) *Nicotiana paniculata* – mesophyll – (Schieder, unpubl.)

Fig. 15. Separation of fusion bodies from the bulk of protoplasts. Illustration of isoosmotic density gradient separation (Harms 1979)

cell wall regeneration in the other cases. These may be properties of the parental protoplasts, influences of the conditions of protoplast isolation or fusion, inappropriateness of the culture conditions, physiological or genetic incompatibilities within the fusion body, or simply that the fusants have not been cultured. It has been previously determined that the degree of relationship of the partners does not play a decisive role; cell walls have even been formed in fusants of *Daucus* protoplasts with cells of *Xenopus* (Davey et al. 1978).

4.2.2 Fates of Nuclei

The behavior of nuclei and chromosomes in fusants has been determined mainly from light and electron microscopical observations on fixed and stained material. During the development of the fusants, the heterokaryotic nature is lost rather early by degeneration or extrusion of nuclei of one type; by segregation particularly as a consequence of nonsynchronized mitosis; or by the formation of hybrid nuclei via fusion of interphase nuclei or of mitotic figures.

4.2.2.1 Homo- and Heterokaryons

Fusion bodies contain only one nucleus in cases of subprotoplast fusion, two nuclei in most of the heterotypic fusion experiments (Fig. 13c), but frequently numerous nuclei (Fig. 3). Heterospecific fusion bodies of *Glycine max* (×) *Nicotiana glauca,* for instance, carried two nuclei to a degree of 45% and three nuclei to 31%, whereas 24% of the heterokaryons had more nuclei (Kao 1977). Multinu-

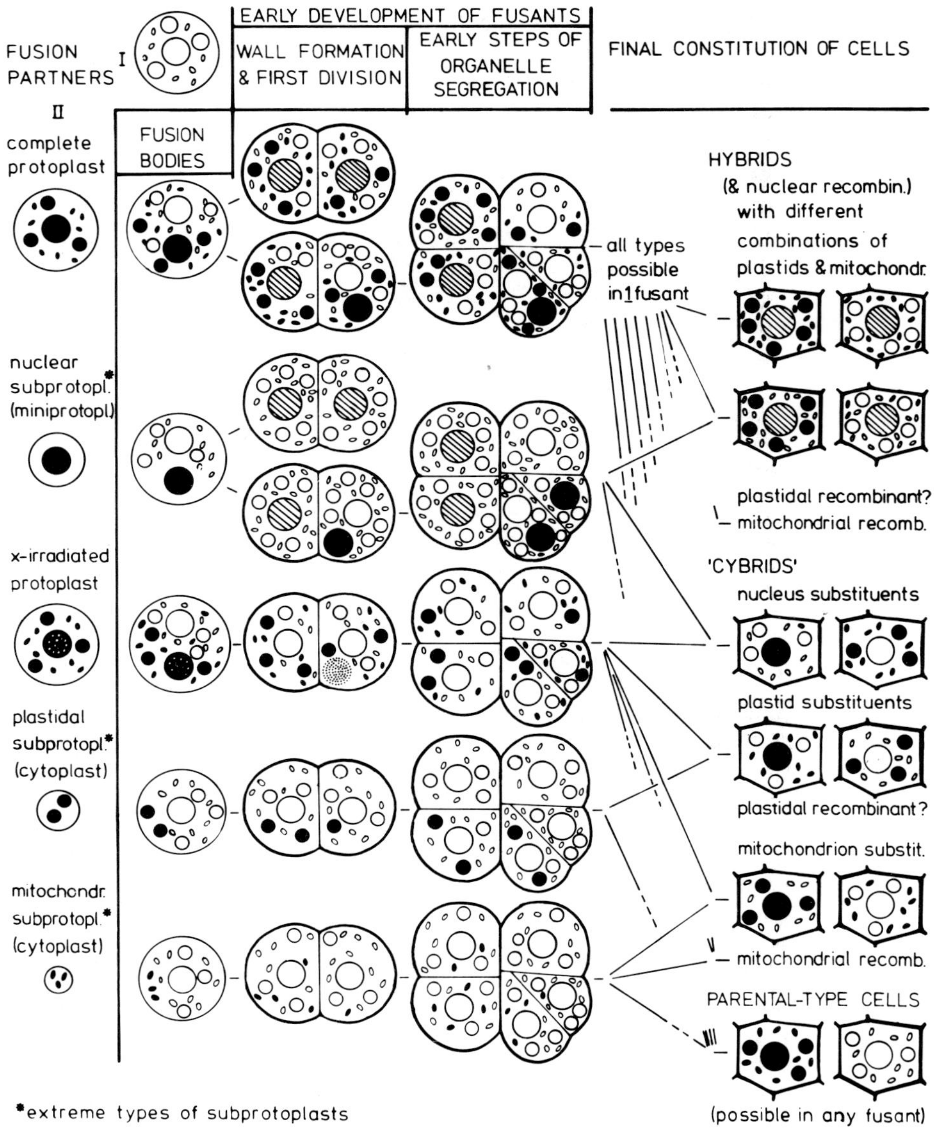

Fig. 16. Possible fates of cell organelles in fusion products of protoplasts and subprotoplasts in early stages of development, as well as after termination of segretion

Table 1. Protoplast fusion experiments in Embryophyta

Systematic relationship and species fused	Most advanced developmental stage	Selected references[a]
Bryophyta		
Hepaticae		
Sphaerocarpaceae		
Sphaerocarpos donnellii (×) *donnellii*	Plantlet	Schieder (1974)
Musci		
Bryaceae		
Bryum erythrocarpum (×) *erythrocarpum*	Fusion body	Binding (1966)
Funariaceae		
Funaria hygrometrica (×) *hygometrica*	Plastocyte	Binding (1966)
Physcomitrella patens (×) *patens*	Plantlet	Grimsley et al. (1977a, b)
Physcomitrium erystomum (×) *erystomum*	Fusion body	Binding (1966)
Spermatophyta		
Magnoliatae		
Apiaceae		
Aegopodium see *Daucus*		
Daucus carota (×) *Aegopodium podagraria*	Plantlet	Dudits et al. (1979)
(×) *capillifolius*	Plant	Dudits et al. (1977) Kameya et al. (1981)*
see also *Nicotiana, Petunia* and *Spinacia*		
(×) *carota*	Plant	Wallin et al. (1974) Harms et al. (1981) Lázár et al. (1981)* Cella et al. (1983)*
(×) *Hordeum vulgare*	Cell cluster	Dudits et al. (1976)
(×) *Nicotiana tabacum*	Cell cluster	Gosch and Reinert (1978)
(×) *Petroselinum hortense*	Plantlet	Dudits et al. (1980)
(×) *Petunia hybrida*	Cell cluster	Reinert and Gosch (1976)
(×) *Vicia faba*	Fusion body	Davey et al. (1980)
Petroselinum see *Daucus*		
Brassicaceae		
Arabidopsis see *Brassica*		
Brassica campestris (×) *Arabidopsis thaliana*	Shoot	Gleba and Hoffmann (1978)
(×) *oleracea*	Plant	Schenk (1982)
(×) *Raphanus sativus*	Plant	Pelletier et al. (1983)
Brassica napus (×) *Glycine max*	Cell cluster	Kartha et al. (1974)
Raphanus see *Brassica*		
Chenopodiaceae		
Spinacia oleracea (×) *Daucus carota*	Plastocyte	Hodgson and Rose (1984)
Fabaceae		
Glycine max (×) *Brassica napus*	Cell cluster	Kartha et al. (1974)
(×) *Colchicum autumnale*	Cell cluster	Constabel et al. (1976)
(×) *max*	Cell cluster	Miller et al. (1971)* Fowke et al. (1975a) Galbraith and Galbraith (1979)
(×) *Hordeum vulgare*	Cell cluster	Kao and Michayluk* (1974)* Kao et al. (1974)*
(×) *Melilotus officinalis*	Cell cluster	Fowke et al. (1976)
(×) *Nicotiana glauca*	Callus	Constabel et al. (1976) Kao (1977)*

Table 1 (continued)

Systematic relationship and species fused	Most advanced developmental stage	Selected references[a]
(×) *Nicotiana langsdorffii*	Cell cluster	Constabel et al. (1976)
(×) *Nicotiana rustica*	Cell cluster	Constabel et al. (1976)
(×) *Nicotiana tabacum*	Callus	Constabel et al. (1976) Chien et al. (1982)*
(×) *Pisum sativum*	Cell cluster	Kao et al. (1974)* Fowke et al. (1977)*
(×) *Vicia hajastana*	Cell cluster	Kao et al. (1974) Constabel et al. (1977)
(×) *Zea mays*	Cell cluster	Kao et al. (1974)
Medicago sativa (×) *falcata*	Plant	Téoulé (1983)
Melilotus see *Glycine*		
Pisum sativum (×) *Nicotiana tabacum* see also *Glycine* and *Vicia*	Fusion body	Syono et al. (1979)
Vicia faba (×) *faba* see also *Daucus* and *Petunia*	Fusion body	Zimmermann and Scheurich (1981)
V. hajastana (×) *Glycine max*	Cell cluster	Kao et al. (1974)* Constabel et al. (1977)*
(×) *Pisum sativum*	Cell cluster	Constabel and Kao (1974)* Kao and Michayluk (1974)* Kao et al. (1974)* Fowke et al. (1975b)
(×) *angustifolia*	Cell cluster	Rennie et al. (1980)
(×) *hajastana*	Cell cluster	Rennie et al. (1980)
(×) *narbonensis*	Cell cluster	Rennie et al. (1980)
(×) *villosa*	Cell cluster	Kao et al. (1974)
Rutaceae		
Citrus sinensis see *Nicotiana*	Callus	Harms and Potrykus (1979) Vardi et al. (1982)
Scrophulariaceae		
Torenia baillonii (×) *Torenia fournieri*	Fusion body	Potrykus (1971)
Solanaceae		
Atropa belladonna see *Datura, Nicotiana* and *Petunia*		
Datura innoxia (×) *Atropa belladonna*	Plantlet	Krumbiegel and Schieder (1979)
(×) *candida*	Plant	Schieder (1980)
(×) *discolor*	Plant	Schieder (1978)
(×) *innoxia*	Plant	Schieder (1977)
(×) *quercifolia*	Shoot	Schieder (1980)
(×) *sanguinea*	Shoot	Schieder (1980)
(×) *stramonium*	Plant	Schieder (1978)
see also *Nicotiana* and *Petunia*		
Hyoscyamus muticus (×) *muticus* see also *Nicotiana*	Shoot	Jia et al. (1983)
Lycopersicon esculentum (×) *peruvianum*	Fusion body	Kinsara and Cocking (1983)
(×) *Nicotiana tabacum*	Plastocyte	Binding (1976)
(×) *Petunia hybrida*	Cell cluster Plant	Binding (1976) Tabaeizadeh et al. (1983)
see also *Solanum*		

Table 1 (continued)

Systematic relationship and species fused	Most advanced developmental stage	Selected references[a]
Nicotiana alata (×) *Petunia parodii*	Fusion body	Patnaik et al. (1982)
N. chinensis (×) *Atropa belladonna*	Shoot	Gleba et al. (1982)
N. debneyi (×) *debneyi*	Plant	Scowcroft and Larkin (1981)
N. glauca (×) *langsdorffii*	Plant	Carlson et al. (1972)*
		Kao et al. (1974)* *
		Smith et al. (1976)*
		Chupeau et al. (1978)*
		Uchimiya et al. (1983)*
see also *N. tabacum* and *Glycine*		
N. langsdorffii see also *Glycine*		
N. plumbaginifolia (×) *plumbaginifolia*	Plant	Marton et al. (1982)*
		Siderow and Maliga (1982)*
		Marton et al. (1983)*
N. rustica (×) *repanda*	Plant	Nagao (1982)
see also *N. tabacum* and *Glycine*		
N. sylvestris (×) *Datura innoxia*	Fusion body	Schieder (1977)
(×) *nudicaulis*	Plant	Galun et al. (1982)
(×) *plumbaginifolia*	Plant	Cséplö et al. (1983)
(×) *repanda*	Plant	Galun et al. (1982)
(×) *rustica*	Plant	Galun et al. (1982)*
		Gleddie et al. (1983)
		Aviv et al. (1984b)*
(×) *sylvestris*	Callus	White and Vasil (1979)
N. tabacum (×) *Atropa belladonna*	Callus	Gosch and Reinert (1978)
		Gleba et al. (1983)*
(×) *Citrus sinensis*	Callus	Harms and Potrykus (1979)
(×) *Datura innoxia*	Callus	Gupta et al. (1982)
(×) *Daucus carota*	Cell cluster	Gosch and Reinert (1978)
(×) *Hyoscyamus muticus*	Callus	Chien et al. (1982)*
		Lázár et al. (1983)*
(×) *alata*	Plant	Nagao (1979)*
		Patnaik et al. (1982)
(×) *glauca*	Plant	Evans et al. (1980)
(×) *glutinosa*	Plant	Nagao (1979)*
		Uchimiya (1982)*
		Horn et al. (1983)*
(×) *knightiana*	Plant	Maliga et al. (1978)*
		Menczel et al. (1981)*
(×) *nesophila*	Plant	Evans et al. (1982)
(×) *otophora*	Plant	Evans et al. (1983)
(×) *paniculata*	Callus	Hein et al. (1982, 1983)
(×) *plumbaginifolia*	Plant	Sidorow et al. (1981)*
		Maliga et al. (1982)*
		Menczel et al. (1983)*
(×) *repanda*	Plant	Nagao (1982)
(×) *rustica*	Plant	Nagao (1978)*
		Iwai et al. (1980, 1981)*
		Douglas et al. (1981)*
		Nakata and Oshima (1982)*

Table 1 (continued)

Systematic relationship and species fused	Most advanced developmental stage	Selected references[a]
(×) *sylvestris*	Plant	Melchers (1977)*
		Zelcer et al. (1978)*
		Aviv and Galun (1980)*
		Medgyesy et al. (1980)*
		Hein et al. (1982)
		Evans et al. (1983)*
		Fluhr et al. (1983)*
		Aviv et al. (1984a)*
(×) *tabacum*	Plant	Withers and Cocking (1972)
		Keller and Melchers (1973)
		Melchers and Labib (1974)*
		Gleba et al. (1975)
		Belliard et al. (1978)*
		Glimelius et al. (1978, 1981)*
		Wullems et al. (1979)*
		Evola (1983)*
		Evola et al. (1983)*
		Grafe and Müller (1983)*
		Koop et al. (1983)*
		Li et al. (1983)*
		Gupta et al. (1984)*
(×) undulata	Plant	Galun et al. (1983)*
see also *Glycine*, *Lycopersicon*, *Petunia*, *Physalis*, *Pisum* and *Salpiglossis*		
Petunia hybrida (×) *Atropa belladonna*	Shoot	Gosch and Reinert (1976, 1978)*
(×) *Datura innoxia*	Fusion body	Schieder (1977)
(×) *Daucus carota*	Cell cluster	Reinert and Gosch (1976)
(×) *Nicotiana tabacum*	Callus	Binding (1976)
		Zenkteler and Melchers (1978)
		Patnaik et al. (1982)
		Steffen and Schieder (1983)*
(×) *axillaris*	Plant	Izhar and Power (1979)
(×) *hybrida*	Plant	Binding (1974a,c)
		Izhar and Power (1979)*
		Bergounioux-Bunisset and Perennes (1980)*
(×) *parodii*	Plant	Power et al. (1976, 1977)*
		Cocking et al. (1977)
		Boeshore et al. (1983)*
		Izhar et al. (1983, 1984)*
(×) *parviflora*	Callus	Berry (1983)
(×) *Vicia faba*	Callus	Binding and Nehls (1978)
see also *Lycopersicon*, *Parthenocissus*, *Scopolia* and *Solanum*		
P. parodii (×) *inflata*	Plant	Cocking (1978)
		Power et al. (1979)
(×) *parviflora*	Plant	Power et al. (1980)
see also *Nicotiana* and *Salpiglossis*		
Physalis minima (×) *Nicotiana tabacum*	Callus	Gupta et al. (1982)

Table 1 (continued)

Systematic relationship and species fused	Most advanced developmental stage	Selected references[a]
Salpiglossis sinuata (×) *Nicotiana tabacum*	Plant (?)	Nagao (1982)
(×) *Petunia parodii*	Callus	Power and Chapman (1983)
Scopolia lucida (×) *Petunia hybrida*	Callus	Power and Chapman (1983)
Solanum nigrum (×) *Lycopersicon esculentum*	Fusion body	Binding and Kollmann (1976)
(×) *Petunia hybrida*	Cell cluster	Nehls (1978)
S. tuberosum (×) *Lycopersicon esculentum*	Plant	Melchers et al. (1978)* Shepard et al. (1983)*
(×) *chacoense*	Plant	Butenko and Kuchko (1980)
(×) *nigrum*	Plant	Binding et al. (1982a)* Gressel et al. (1984)*
(×) *stenotum*	Cell cluster	Hein et al. (1982)
Vitaceae		
Parthenocissus tricuspidata (×) *Petunia hybrida*	Callus	Power et al. (1975)
Liliatae		
Liliaceae		
Allium cepa (×) *cepa*	Fusion body	Bracha and Sher (1981)
Colchicum autumnale see *Glycine*		
Lilium longiflorum (×) *Trillium kamtschaticum*	Fusion body	Ito and Maeda (1973)
Poaceae		
Avena sativa (×) *sativa*	Fusion body	Withers and Cocking (1972)*
Hordeum vulgare see *Daucus* and *Glycine*		
Zea mays (×) *Triticum aestivum* see also *Glycine*	Plastocyte	Harms and Potrykus (1978)

[a] References are marked by * when the indicated developmental stage was obtained.

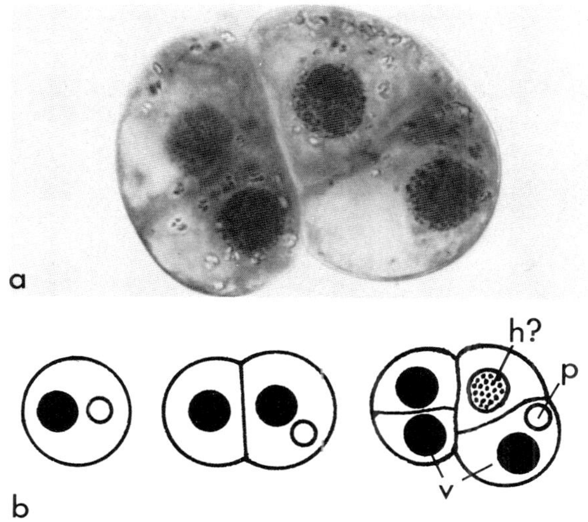

Fig. 17 a, b. Early steps of segregation. **a** *Vicia faba* (×) *Petunia hybrida;* **b** reconstruction of the probable steps of development; hybrid nature of one nucleus (*h*) is concluded only by its two nucleoli of different sizes; (*v*) nucleus of *Vicia*, (*p*) nucleus of *Petunia* (Binding and Nehls, unpubl.)

cleate fusion bodies arise either from multiply-induced fusion or from the participation of spontaneous fusion products in induced fusion. Highly polyenergidic protoplasts showed restricted regeneration. In soybean (×) pea heterokaryons, plastocytes were formed, but did not divide; the nuclei were highly lobed and chloroplasts and many tiny vacuoles were clustered around them (Fowke et al. 1977). Multinucleate protoplasts of various combinations were found to deteriorate (Gamborg et al. 1981). The nuclei of recently fused protoplasts are distributed at random. During the first hours of culture, they frequently assemble (e.g., Miller et al. 1971; Gosch and Reinert 1976, 1978; Binding and Nehls 1978).

The heterokaryotic nature may be transmitted to the daughter cells. This is most reliably established when mitosis occurs unsynchronized, as it has been observed in spontaneous fusants of soybean (Miller et al. 1971) and in heterokaryocytes of *Vicia faba* (×) *Petunia hybrida* (Fig. 17; Binding and Nehls 1978; Nehls and Binding 1979) and of *Solanum nigrum* (×) *Petunia hybrida* (Fig. 18; Nehls 1978). The formation of heterokaryotic daughter cells has also been observed after more or less good synchronization of mitoses, accompanied by multiple wall formation (Kao et al. 1974; Constabel et al. 1977; Gosch and Reinert 1978).

4.2.2.2 Cybrid Formation

In accordance with Gleba and Sytnik (1984), it is proposed to apply the term cybrid to cells and organisms which carry an uniparental nucleus associated with foreign extrakaryotic genetic traits, irrespective of whether the plasmon is composed entirely of uniparental genophores or whether it represents heterogenetic or recombinant genophores. The term was introduced to plant cell fusion genetics by Cocking (1977) and was limited to fusants of a cytoplast to a whole cell by the terminology committee of the Tissue Culture Association (1984; see Appendix, 2). However, there exist several pathways leading to a cybrid configuration.

Direct cybridization is achieved when cytoplasts are fused to protoplasts (Binding 1976) or to miniprotoplasts, or when X-irradiated protoplasts are used (see Sect. 3.4.3). Cybrid cells are also formed during the early development of heterokaryotic fusants. A possible process is the elimination of a nucleus by budding as it has been observed in uniparental plastocytes (Miller et al. 1971; Binding and Kollmann 1976). Another way is verified by nucleus segregation (Figs. 17, 18). This is, for instance, established by asynchronous mitosis leading to a heterokaryotic daughter cell – as it has been mentioned in Sect. 4.2.2.1 – and a cell containing solely the daughter nucleus.

In extreme cases, one of the parental nuclei in the fusion bodies never divides. This is true in cases of inactivation of the nuclei by X-rays as introduced by Zelcer et al. (1978). It is also supposed to be the case if a nucleus is prevented from division by factors of its environment, e.g., by incongruity of metabolism in far-related combinations, by incompatibility as suggested for fusants *Petunia* (×) tomato (Binding 1976), or by inappropriate composition of the culture media. However, no clear-cut evidence as to whether actually one or the other case occurred or if it is even possible in certain combinations is available at present.

In the cases in which cybrid plants have been recovered from protoplast-to-protoplast fusion experiments (e.g., in *Nicotiana tabacum*, Gleba 1979; in *Sola-*

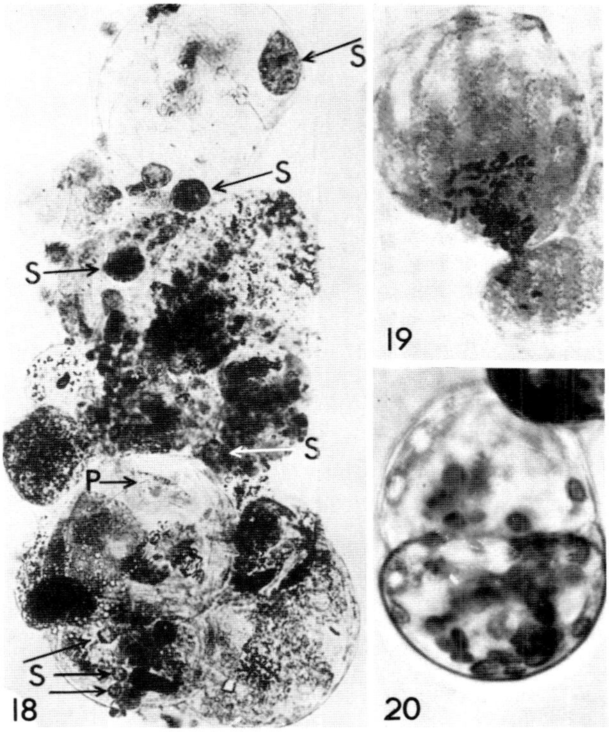

Fig. 18. Early steps of segregation. *Solanum nigrum* (×) *Petunia hybrida* cell cluster grown selectively after blocking of parental protoplasts by diethylpyrocarbonate and iodoacetate, respectively; (*S*) nuclei of *Solanum* type, (*P*) of *Petunia* type (Nehls 1978)

Fig. 19. Early steps of segregation. Protoplast of soybean losing chromosomes by budding (Miller et al. 1971)

Fig. 20. Early steps of segregation. *Vicia faba* (×) *Petunia hybrida;* unequal distribution of the *Petunia* chloroplasts (Binding and Nehls 1978)

num nigrum (×) *tuberosum,* Binding et al. 1982a), it cannot be decided if they arose in reality by a protoplast-to-cytoplast (see Sect. 3.4.4.1) or by one of the described developmental events in heterokaryons.

4.2.2.3 *Hybrid Formation*

Nuclei which were apparently composed of subunits representing original parental nuclei have been found in fixed preparations (Miller et al. 1971; Kao et al. 1974; Constabel et al. 1975a; Dudits et al. 1976; Fowke et al. 1977; Kao 1977; Binding and Nehls 1978). Additionally, Miller et al. (1971) attributed polyploid mitoses in plastocytes which were derived from cell suspensions of soybean with stable, lower chromosome numbers to previous nuclear fusion. Fowke et al. (1977) detected bridges between pea and soybean nuclei in electron micrographs.

These channels, probably, represented an initial step of interphase nucleus fusion.

It has been more frequently observed that nuclear material was combined during mitosis of plastocytes (e.g., Kao et al. 1974; Kao 1977; Gosch and Reinert 1978; Chien et al. 1982). Fairly synchronized mitosis is demanded for this pathway, whereas – vice versa – a hybrid nucleus is not an obligatory consequence (see Sect. 4.2.2.1). Synchronization has been observed even in combinations of suspension culture cells in the G_1 phase and mesophyll cell in the G_0 phase. It appeared that the time of mitosis in heterokaryotic plastocytes was delayed in comparison to pure cell suspension plastocytes which were the faster parent (Constabel et al. 1976). No clear information is available on the process of synchronization. It has been discussed that close contact or even bridges between the nuclei must exist as they have been suggested to be also involved in nucleus fusion (Fowke et al. 1975 a).

All observations on the fusion of mitotic figures have been made in plastocytes. As mentioned in Sect. 4.2.2.1, heterokaryotic daughter cells do occur and, hence, delayed formation of hybrid nuclei may be expected, giving rise to a hybrid/cybrid mosaic which has probably occurred in the cell cluster shown in Fig. 17.

4.2.2.4 Fates of chromosomes

The feature of karyotypic changes in developing fusants is dealt with in Part III of this volume. However, it is interesting to know that observations in mitotically active plastocytes already revealed some information on the probable mechanisms of cytogenetic variation. Miller et al. (1971) were able to visualize metaphase chromosomes being separated from the others by budding in a soybean homoplastocyte (Fig. 19). Chromosomes were found in the cytoplasm of daughter cells of a homoplastocyte of *Vicia faba* (Binding and Nehls 1978). Kao (1977) described incompletely synchronized mitoses of *Glycine max* (×) *Nicotiana glauca* heteroplastocytes already showing various chromosomal aberrations. Corresponding observations were found in the second mitosis of *Glycine max* (×) *Nicotiana tabacum* by Chien et al. (1982).

4.2.3 Fates of Plastids

Plastid segregation and incompatibility in the early development of fusants has been discussed repeatedly (e.g., Binding and Kollmann 1976; Cocking 1977; Binding 1979). Unequal distribution of the different plastids to the daughter cells after the first cytokinesis has been found in protoplast-to-protoplast fusants of *Vicia faba* (×) *Petunia hybrida* and in protoplast-to-subprotoplast fusants of *Petunia hybrida* (×) *Lycopersicon esculentum* (Binding 1976). Their random location in the cytoplasm or incomplete mixture of the protoplasts might have been responsible for the unequal distribution. Peculiar development of plastids in fusants of *Petunia hybrida* (×) *Nicotiana tabacum* (Binding 1976) has been attributed to incompatibility. Tobacco chloroplasts were found in extremely divergent numbers in the cells of small cell clusters and some had formed large starch grains.

Hodgson and Rose (1984) observed degenerating chloroplasts in spinach mesophyll (x) carrot root parenchyma heteroplastocytes. The simultaneous presence of chloroplasts with integer ultrastructure was explained by different organization forms. Incompatibility has been considered as a possible explanation also for the degeneration of chloroplasts in homoioplasmic fusants of cell culture and mesophyll protoplasts suggesting that incompatibility-like reactions may also be caused by the interference of differently developed organelles (Fowke et al. 1975a, 1976).

4.2.4 Formation of Genetic Mosaics

As illustrated in Fig. 16 and discussed in the preceding chapters, events during the development of fusants may lead very early to mosaics of variant cells. Naturally, the diversity is much lower in intraspecific than in interspecific fusants. Localization of certain types of cells and competition within the regenerants may decide on transmission or elimination of particular recombinants. It is, therefore, helpful in interspecific combinations to make use of a selective pressure to a desired cell type or to separate the cells as early as possible. Subcloning has been established by reisolation of protoplasts (Binding et al. 1982a).

Segregation of nuclei, leading to cybrid cells, or formation of hybrid nuclei are, certainly, completed after a few cell divisions. Segregation, loss and rearrangements of chromosomes, plastids, and mitochondria may be extended to later stages of development up to sexual progenies. These features are covered in the following part of this volume. Phenomena of incompatibility and incongruity will also be discussed in more detail in Part III.

5 Conclusions and Aspects on Differentiation of Fusants in Early Development

Detailed observations on the early development of fusion products are relatively rare as compared to the period of more than 10 years since the establishment of protocols for highly efficient protoplast fusion. Nevertheless, it became evident that the genetic nature of the fusant clones is most highly dependent not only on the compositions of the parental protoplasts or subprotoplasts, but also decisively on the fates of cell organelles in the first or next following divisions of the fusant. Above all, it is decided in that period whether only hybrid cell lines or cybrids or hybrid/cybrid mosaics arise.

The clonal variation of fusion products is a means for the achievement of high genetic variability, including cell organelle substitution products (Fig. 16) and various types of genetic recombinants which have – with poor success so far – also attempted the assessment by organelle transplantation and – with increasing success – by DNA transfer.

Differentiation of fusants will be discussed in more detail in the following parts of this volume when experimental results obtained with advanced developmental stages are surveyed.

Acknowledgements. The authors wish to thank various colleagues for providing photographs and for permission to use their figures for illustration.

References: See end of Chapter III

III Development of Protoplast Fusion Products

R. NEHLS [1], G. KRUMBIEGEL-SCHROEREN [2], and H. BINDING [2]

1 Introduction

The preceding chapter was devoted to the establishment of protoplast fusion bodies and processes of their early development; growth and differentiation of fusant cell clones and plants are treated in this part of the Volume. Proliferation and organogenesis, segregation and recombination of plastids and mitochondria, as well as loss of chromosomes, are prominent features now being investigated with respect to competition between interspecific genetic traits, to the genesis of somaclonal variation, and to the construction of new genotypes.

An increasing number of intrageneric combinations has been and is being investigated. Comparably few combinations have been established between species of lower degrees of relationship. Plants of economic interest have frequently been used in somatic cell hybridization experiments (e.g., soybean: Kao 1977; potato: Melchers et al. 1978; Butenko and Kuchko 1979, 1980; Komarnitzky et al. 1980, 1981; Binding et al. 1982a; Barsby et al. 1984; Gressel et al. 1984; rape: Pelletier et al. 1983), and successful utilization of somatic hybrids in the establishment of new cultivars is in progress.

It is not intended here to accumulate all publications so far available, but to quote only selected papers. Some additional references are cited by Gleba and Sytnik (1984). Combinations of species and developmental stages obtained with fusion products are compiled in Table 1 of Chap. II of this Volume. Growth conditions for clonal propagation, induction of organogenesis, and regeneration of plants are basicly the same as described for uniparental cultures (Chap. I, this Vol.). A few modifications have been applied to hybridization in cases in which certain environmental factors could be utilized for selection (see Sect. 2.2).

2 Genetic Traits Utilized in Somatic Hybridization Experiments

The populations resulting from a fusion experiment are usually composed of the heterotypic fusants, homotypic fusants, and monotypic parental individuals

[1] PLANTA Angewandte Pflanzengenetik und Biotechnologie GmbH, Postfach 146, 3352 Einbeck, FRG.
[2] Botanisches Institut der Christian-Albrechts-Universität, Biologiezentrum, Olshausenstraße 40–60, 2300 Kiel 1, FRG.

Results and Problems in Cell Differentiation 12
Differentiation of Protoplasts and of Transformed Plant Cells (Edited by J. Reinert and H. Binding)
© Springer-Verlag Berlin Heidelberg 1986

– as far as single pairs of protoplasts were not treated. The genetic variability is further increased by somaclonal variation within the fusant clones as outlined in Fig. 16 of Chapter II. Hence, it is evident that great efforts have been made to elaborate marker systems and techniques for selection, recognition, identification, characterization, and analysis of the fusants. Whereas the investigations with intraspecific combinations relied on bi- or oligofactorial traits, more versatile and complex possibilities are at the investigator's disposal in hybridization of less related species. For early recognition and selection of fusants, the reader ist referred to Chapter II, Sections 3.4.3 and 4.1. Whereas in early developmental stages, differentiation of the protoplast donor cells could be utilized, differential properties of more developed fusants are dependent solely on genetic traits.

2.1 Differential Growth Requirements

Differential growth requirements have mainly been employed in selecting fusion products. Either one or even both parents could be eliminated by the inability to grow under the applied culture conditions. In the first attempts to regenerate somatic hybrids of higher plants, Carlson et al. (1972) made use of differential demands for regeneration of the parental *Nicotiana* species and their sexual hybrid. This knowledge was utilized in other experiments in this genus (Smith et al. 1976; Maliga et al. 1977; Chupeau et al. 1978; Uchimiya et al. 1983) and in *Petunia* (Power et al. 1977). In other cases, selective growth of the hybrids was expected from particular properties of the parental types or it happened fortunately to appear in the course of the experiment.

2.1.1 Differential Responses to Unspecified Factors of Culture Media

Selective pressures by more or less undefined culture conditions were observed in several systems. In the combination *Arabidopsis thaliana* (×) *Brassica campestris* cell divisions sustained only in the hybrid cell lines (Gleba and Hoffmann 1978). The parental protoplasts of *Brassica campestris* and *B. oleracea* died after producing dark-brown particles, whereas the fusion products – the resynthesized rapeseed – proliferated (Schenk and Röbbelen 1982; Schenk 1983). Selection of *Nicotiana* interspecific hybrids relied also partly on differential media demands (Carlson et al. 1972). Heterosis-like growth of fusant calluses may also be mentioned in this context (Schieder 1977; for further examples see Sect. 4 and 8.4). Unilateral elimination of one parental strain by unknown experimental factors helped to select fusion products in several other cases (e.g., Power et al. 1975, 1979, 1980; Evans et al. 1980; Binding et al. 1982a).

2.1.2 Differential Wild-Type Characters in the Response to Distinct Culture Conditions

Better defined cultural factors could also be applied to fusant selection: Increased phytohormone production of the *Nicotiana glauca* (×) *langsdorffii* hybrids made possible their selective growth on hormone-free media (Carlson et al. 1972; Smith et al. 1976). Temperature in combination with certain compounds of the culture media were selective against parental *Solanum* species (Shepard et al. 1983). Zelcer et al. (1978) identified mannitol as a selective agent in the genus *Nicotiana*.

2.1.3 Auxotroph Complementation Systems

Several investigations have been undertaken in order to construct dihybrid complementation systems. Utilization of auxotrophs in hybridization of plants was successful nearly simultaneously in fungi (Ferenczy et al. 1974; Binding and Weber 1974) and in the liverwort *Spaerocarpos donellii* (Schieder 1974). Later on, it was used in the moss *Physcomitrella patens* (Ashton et al. 1977 a, b) and in angiosperms (Sidorov and Maliga 1982; Fankhauser et al. 1983; Jia et al. 1983; Shimamoto and King 1983; Potrykus et al. 1984). A tobacco mutant affecting the utilization of glycerol (clone Gut, Chaleff and Parsons 1978 b) was incorporated into a selection system by Evola et al. (1983).

The isolation of nitrate reductase-deficient mutants (NR^-) in dihaploid tobacco cell cultures by Müller and Grafe (1968) laid the basis for the establishment of complementation selection as elegantly as is possible in mammalian cell fusion by the HAT technique (Littlefield 1974). NR^- mutants can be selected by their resistance to chlorate, which is reduced by the functional NR-enzyme to the toxic chlorite. Parental NR^- cell lines are eliminated in hybridization experiments by their inability to utilize nitrate which is incorporated into the culture media as the only nitrogen source. For fusion, two different mutant types are combined: the nia type which is deficient in the functional apoenzyme, and the cnx type which lacks the molybdenum cofactor (Glimelius et al. 1978; Grafe and Müller 1983; Xuan et al. 1983). NR^- mutants of other species were also used in hybridization experiments (*Hyoscyamus niger:* Fankhauser et al. 1983; Lázár et al. 1983; *Nicotiana plumbaginifolia:* Marton et al. 1982; *Petunia hybrida:* Steffen and Schieder 1983, 1984). The suitability of NR^- mutants for selection was also demonstrated in miniprotoplast fusion (Wallin et al. 1979), in intergeneric gene transfer mediated by X-irradiated protoplasts (Gupta et al. 1982), and by utilizing mutant cells in nurse culture (Hein et al. 1983).

2.2 Differential Tolerance / Resistance

2.2.1 Sensitivity to Light

A hybrid complementation system of recessive sensitivities to high light intensities was devised and emphasized as being widely applicable to plant fusant selection by Melchers and Labib (1974). Two chlorophyll-deficient, light-sensitive tobacco mutants were combined to fusants expressing normal high light resistance.

2.2.2 Response to Drug and Disease

Power et al. (1976, 1977) discovered naturally occurring differences in actinomycin D tolerance in different *Petunia* species which were useful in hybrid selection. It was supposed that such differences may also occur in other taxa; but this has apparently not been investigated so far. Drug-resistant mutants were usually isolated by the application of sublethal drug concentrations and, after some growth, of completely selective concentrations (Binding et al. 1970). Because of the comparably easy isolation of drug-resistant mutants, Maliga et al. (1977) stressed the suitability of double resistance for hybrid selection.

Resistance has been found, mainly by somatic hybridization, to be dominant, recessive or intermediate, nucleus-coded, or under the control of extrakaryotic genophores. Resistances to actinomycin D (Power et al. 1976, 1977) and cycloheximide (Lázár et al. 1981) were expressed recessively. Dominant expression of drug resistances in intraspecific somatic hybrids was observed with DL-5-methyltryptophan in *Daucus carota* (Harms et al. 1981, 1982), and *Nicotiana tabacum* (Horn et al. 1983), with S-(2-aminoethyl)-L-cystein (AEC) in *N. sylvestris* (White and Vasil 1979) and in *Daucus carota* (Harms et al. 1981), with algo-azetidine-2-carboxylic acid (A2 CA) in *D. carota* (Harms et al. 1981; Cella et al. 1983), with picloram and with hydroxyurea in *N. tabacum* (Evola et al. 1983). Resistances were fully expressed also in the interspecific somatic hybrids *D. carota*-5-methyltryptophan^r/A2CA^r (×) *capillifolius*-wild-type (Kameya et al. 1981), *N. tabacum*-5-methyltryptophan^r (×) *D. carota*-AEC^r (Hauptmann et al. 1983), and *Nicotiana tabacum*-ACE^r (×) Daucus carota-5-methyltryptophan^r/A2CA^r (Harms et Oertli 1982). Resistance to 5-methyltryptophan was reduced in hybrids of *Nicotiana tabacum* (Gleba and Berlin 1979; Gleba 1980) and of *Nicotiana sylvestris* (White and Vasil 1979).

Cytoplasm-inherited resistances have been utilized for selection of fusants and as tracers for plastids. They were applied to intraspecific and interspecific fusion in *Nicotiana, Petunia, Solanum,* and *Brassica.* Selective drugs were kanamycin (Maliga et al. 1977), tentoxin (Aviv and Galun 1980; Aviv et al. 1980; Galun et al. 1982a, b), streptomycin (Medgyesy et al. 1980, 1983; Menczel et al. 1983), lincomycin (Cséplö et al. 1983) and the herbicide atrazin (Binding et al. 1982a; Pelletier et al. 1983; Gressel et al. 1984; Robertson and Earle 1985). Herbicide-resistant biotypes have already been found in a number of species (cf. Le Baron and Gressel 1982). As they are only expressed after chloroplasts are differentiated, herbicides are used not as selective agents but for identification of the plastids in fusant clones.

Mitochondria-located resistance could so far not be utilized in fusant analysis.

Resistance to tobacco mosaic virus (TMV) was expressed in somatic hybrids of sensitive tobacco and resistant *Nicotiana glutinosa* (Uchimiya 1982), *N. repanda* (Nagao 1982), and *N. nesophila* (Evans et al. 1981).

2.3 Differential Pigmentation

Chlorophyll-deficient mutants were frequently utilized as markers for recognition of fusants. They appear at relatively high frequencies in seedling populations and as somaclonal variants and are easily selected. They are available already in a wide range of species. Alternate complementation of nonallelic chlorophyll deficiencies was obtained, for instance, in *Petunia* (Binding 1974c; Berry 1983), in *Nicotiana* (Melchers and Labib 1974, Douglas et al. 1981a; Sidorov and Maliga 1982), and *Datura* (Schieder 1977). Discrimination of one parental type was possible in fusants of chlorophyll-deficient and green parents (Schieder 1974, 1978, 1980a; Cocking et al. 1977; Melchers et al. 1978; Dudits et al. 1979, 1980; Krumbiegel and Schieder 1979, 1981; Power et al. 1979, 1980; Evans et al. 1980; Berry 1983; Power and Chapman 1983).

Intermediate levels of chlorophyll deficiency were appropriate for selection of somatic hybrids heterozygous in the sulfur gene (Su/su). Twin spots on leaves following somatic crossing-over was taken as an additional indication for the hybrid nature of the regenerated plants (for ref. of Gleba and coworkers see Gleba and Sytnik 1984; Evans et al. 1980, 1981, 1983).

Further nucleus-coded color markers which helped recognition of hybrids were, for instance, anthocyanin pigmentation of stems and leaves in *Nicotiana* (Evans et al. 1981) and in *Solanum* (Shepard et al. 1983), as well as carotenoid formation in the roots of Apiaceae (Dudits et al. 1979).

Plastome albino mutants are useful for fusant selection and highly qualified as tracers for the plastids. They have been used for instance, in *Nicotiana* (Gleba and coworkers, see Gleba and Sytnik 1984; Glimelius and Bonnett 1981; Sidorov et al. 1981). It is well known from interspecific sexual hybrids that chlorophyll deficiencies (bastard bleaching) originate by certain genome/plasmone interactions (see Sect. 8.1). Correspondingly, light green pigmentation was found and used for selection of a cybrid in *Solanum nigrum* (×) *tuberosum* (Binding et al. 1982a).

2.4 Differential Morphology

With decreasing relationship of the fusion partners, increasing numbers of morphological properties can be utilized Only some examples are mentioned in the following section: Trichomes were particularly appropriate in the systems *Arabidopsis* (×) *Brassica* (absence or presence of trichomes and their shapes; Gleba and Hoffmann 1979, 1980), *Datura* (×) *Atropa* (presence or absence of hairs on callus; Krumbiegel and Schieder 1979), *Solanum nigrum* (×) *tuberosum* (length of gland hairs in shoot culture, hair types at calyx; Binding et al. 1982a), *Nicotiana glauca* (×) *tabacum* (hair density at lower leaf epiderm; Evans et al. 1980) and *Nicotiana tabacum* (×) *nesophila* (same character as former system; Evans et al. 1982).

Additional useful characters were the sizes and shapes of leaves (Gleba and Hoffmann 1979; Glimelius and Bonnett 1981; Glimelius et al. 1981; Binding et al. 1982a; Tabaeizadeh et al. 1983). Investigated flower properties implied size and shape of calyx and corolla, flower tube diameter, limb width (Evans et al. 1980, 1982; Glimelius et al. 1981; Binding et al. 1982a), and single or branched arrangement of various flower types (Gleba and Hoffmann 1979, 1980).

2.5 Differential Organogenetic Potential and Hybrid Malformations

Selective capability of hybrid clones to form shoots has been described by Maliga et al. (1977) and Power et al. (1979, 1980), whereas in some interspecific and intergeneric hybrids reduced organogenetic potency was stepwise recovered and malformations were overcome (see Sects. 5.2.1 and 8.5.2.2).

Disturbed development of flower organs allowed a conclusion on their impaired functionality (Gleba and Hoffmann 1979, 1980). Abnormal development of stamina and stigma were significant symptoms, indicating the fates of mitochondria when one of the parents carried cytoplasmic male sterility (see Sects. 5.7 and 8.5.2.5).

2.6 Cytological Properties

In most fusion combinations, investigation of the karyotypes was an essential part of the methodological confirmation of the hybrids and investigation of their development. Especially in fusion products of near related species, only chromosome numbers could be used for discrimination between parental and fusant clones. It is obvious, however, that increased numbers are weak indications for heterotypic fusants, as spontaneous and homotypic fusants, as well as autopolyploids, must be taken into account. As an exception, clearcut proof by chromosomal constitution of the hybrid nature was possible even in an intraspecific combination: Schieder (1974) chose male and female parents of *Sphaerocarpos donnellii* to identify the hybrids by their content of both types of gonosome. In most of the intergeneric fusion experiments, the partners were distinguished in chromosome numbers as well as sizes and shapes (e.g., Power et al. 1975; Gosch and Reinert 1976, 1978; Constabel et al. 1977; Kao 1977; Binding and Nehls 1978, 1979; Krumbiegel and Schieder 1979; Wetter and Kao 1980; Chien et al. 1982; Gleba et al. 1982; Tabaeizadeh 1983), or they were at least clearly discernable by some marker chromosomes (Gleba and Hoffmann 1978; Hoffmann and Adachi 1981). Nuclei could be identified by characteristic chromocenters (Maliga et al. 1978; Nehls 1978). Electron microscopic morphometry of chromatin was carried out in *Arabidopsis* (×) *Brassica* by Nagl and Hoffmann (1980).

As an additional marker of cell structure Scowcroft and Larkin (1981) evaluated the numbers of plastids in guard cells in order to draw conclusions on the ploidy level of the putative hybrids (cf. Butterfaß 1964). It is evident from electron microscopic investigation of graftings that a number of further ultrastructural marker systems should be well suited for the examination of fusants (Kollmann and Glockmann 1985).

2.7 Differential Structure of DNA

2.7.1 Investigation by DNA-DNA Hybridization

The comparably simple technique of isolating total DNA from a putative somatic hybrid and studying the degree of molecular reassociation with parental DNA may be useful to calculate the amount of genetic information of one or the other parent in hybrids in which the chromosomes are insufficiently distinguishable. Dudits et al. used uniparental total DNA as reference (1979). A cloned genome bank was applied by Saul and Potrykus (1983, 1984).

Since fractionating DNA by means of restriction endonucleases has become a routine technique in recent years (see Sect. 2.7.2), it has also been used in combination with fragment hybridization. The identification of particular restriction bands by nick-translated probes gives significant information on the constitution of fusants (Boeshore et al. 1983; Galun et al. 1983; Müller-Gensert et al. 1983; Uchimiya et al. 1983, 1984; Brar et al. 1984). Furthermore, it offers the opportunity to investigate DNA from all the cellular genophores in a single DNA sample, assuming that the probes are sufficiently sensitive and cross-hybridization does not occur in misleading amounts (Müller-Gensert et al. 1983).

2.7.2 Analysis by Restriction Endonuclease Fragmentation

Differently from the use of DNA fragments in molecular hybridization analysis, the patterns of the fragments themselves can be taken as a means of distinguishing between parental and hybrid characteristics. The large and still increasing number of restriction endonucleases available enables specific fractionation of DNA, and thus the detection of even minor differences in their structure. Especially plastid and mitochondrial DNA with sizes much smaller than nuclear DNA are suited for analysis by comparison of restriction patterns. It has therefore become possible to follow up the fates of organelles in somatic hybrids and their segregation in the mitotic progeny as well as to find evidence for or against molecular recombination following protoplast fusion. To obtain significant restriction patterns of single types of organelle, very pure DNA preparation must be used. Contamination by DNA of other origin must be completely avoided. This fact reflects the high sensitivity of the method which allows the detection of less than 0.1% of homology to parental DNA (Scowcroft and Larkin 1981; Schiller et al. 1982). The technique was applied, for instance, by Chen et al. (1977), Komarnitzky et al. (1981), Nagy et al. (1981), Galun et al. (1982b), Aviv et al (1984), Fluhr et al. (1984), Pelletier and Chupeau (1984), and Chetrit et al. (1985).

2.8 Differential Protein Patterns

2.8.1 Analysis of Protein Patterns Irrespective of Their Functions

Preparations of extracted total soluble protein or specific fractions, as well as from mitochondria and plastids, can be submitted to electrophoresis or electrofocusing.

The information which can be drawn from such experiments is mainly dependent on the existence of major distinguishable markers in the protein patterns of the parents. It must be ensured that alterations due to different physiological states do not occur with such markers or that they can be recognized. Furthermore, the risk for detection of additional fractions from degradation of labile proteins must be taken into consideration, as with any other technique that relies on electrofocusing.

Cautiously handled, patterns of protein extracts offer the opportunity of screening for hybrid characters in systems lacking selectable traits.

2.8.2 Isoenzyme Analysis

Differences in electrophoretic mobilities of enzymes catalyzing the same biochemical processes in vitro can be utilized in order to distinguish between traits that are expressed in tissue regenerated after protoplast fusion. The techniques are mostly quite simple and quickly done, but the results may sometimes be embarrassing. This is true especially with enzymes expressing peroxydase activity, which is easily registered after electrophoresis. The control of its expression in different plant organs and under diverse physiological conditions, however, is not understood. Even standardization of age and growth condition did not completely overcome these problems (Scandalios 1974). Peroxidase isoenzymes have nonetheless been used successfully in a number of experiments to detect the hybrid nature of protoclones (Carlson et al. 1972; Power et al. 1975; Dudits et al. 1977; Gleba and Hoffmann 1978).

There are further enzymes that were useful to analyze fusants and of which the genetic organization of expression is partly known. Among these, malate dehydrogenase (Wetter and Kao 1976), α-amylase (Lönnendonker and Schieder 1980), lactate dehydrogenase (Wetter and Kao 1976; Maliga et al. 1977), esterases (Wetter and Kao 1976; Maliga et al. 1977; Menczel et al. 1978; Medgyesy et al. 1980; Sidorov et al. 1981; Maliga et al. 1982) and others.

In some investigations, additional bands of isoenzymes exhibiting electrophoretic properties different from those in the parental cells have been found. These could be regarded as either artifacts (Leible et al. 1982) or hybrid molecules, but have also been interpreted as the result of derepression of silent genes already present in the parental cells. Thus the genes are expressed in the new genetic environment (Gleba and Hoffmann 1978).

Isoenzymes showing an electrophoretic mobility intermediate between those of the parental molecules were also regarded as artifacts by some authors (e.g., O'Connell and Brady 1981), arguing that such variations are often observed in isoelectrofocusing. Most enzymes are Mendelian inherited, so isoenzyme analyses are mainly useful for the investigation of nuclear traits. One important exception is ribulose-1,5-bisphosphate carboxylase/oxygenase (see Sect. 2.8.3).

2.8.3 Fraction-I-Protein

The fraction-I-protein, exhibiting both the ribulose-1,5-bisphosphate carboxylase and oxygenase activities, is the most abundant protein complex in the plant. It represents about 40% of the total soluble leaf protein. In tobacco, it can be isolated by direct crystallization (Chen et al. 1972) but this is not possible in other species.

Fraction-I-protein has a unique property which makes it an ideal marker to follow up both nuclear and chloroplastic traits after somatic hybridization. The protein is located within the chloroplasts and is composed of 16 subunits with an overall molecular weight of 560,000. Eight subunits are of a larger type (MW 55,000) and are coded for by chloroplast genes (v. Wettstein et al. 1978). The other eight subunits belong to the smaller class (MW 12,000–14,000) and are coded for by nuclear genes.

In species other than tobacco, fraction-I-protein is isolated by immunological methods (Gray and Wildman 1976), and the subunits submitted to isoelectrofocusing.

The pattern of subunits with different isoelectric points is an excellent marker to examine nuclear and chloroplastic traits at the same time. It has been widely used to confirm the hybrid nature of regenerants and to follow up the segregation of specific characters (e.g., Melchers et al. 1978; Zelcer et al. 1978; Poulsen et al. 1980; Iwai et al. 1980; Komarnitzky et al. 1980; Douglas et al. 1981 b; Glimelius et al. 1981).

Aberrations, additional bands, and hybrid bands are also found with electrofocusing of fraction-I-protein. Reasons for such phenomena are being sought as with other isoenzymes (see Sect. 2.8.2).

2.9 Differential Low Molecular Compounds

Ninnemann and Jüttner (1981) introduced a technique for the examination of hybrid cell progenies, which was new for this purpose and has to date remained unique. They investigated low molecular compounds from plants whose hybrid nature had already been confirmed. Volatile substances could be separated from each other and identified as inherited by one of the parents or as being new by gas chromatography.

The problems concerned with this method are similar to those with the analysis of isoenzymes: it might be difficult to explain the occurrence or absence of a compound only as an expression of hybrid or nonhybrid state. The results obtained may also reflect different physiological activities.

3 Isolation and Identification of Fusants and Their Clonal Variants

3.1 Selection and Identification

The choice of marker systems for the seizure of fusant clones and variants depends on the availability of the markers, and on the expected degree of somaclonal variability, or on the desired recombinant. Fusants have usually been selected by the use of a single pair of markers of intermediate expression in the hybrids, or by the use of bifactorial combinations leading either to complementation or to bilateral expression.

For unequivocal proof of the fusant nature, it must be excluded that the presence of biparental traits is based on mosaic constitution of the regenerant. The risk of misinterpretation was most strikingly demonstrated by Brabec (1954) when he disproved the hybrid nature ("burdo" character) of isolates grown from *Lycopersicon esculentum* (×) *Solanum nigrum* graftings by Winkler (1938). In somatic hybridization experiments, for instance, cross-feeding may simulate hybrid auxotroph complementation; and hybrid patterns of proteins (Sect. 2.7) or DNA fragments (Sect. 2.8) may also be indistinguishable from those derived from a mixture of different cell lines.

Discrimination between fusant and mosaic is possible by the separation of single cell lines (Sect. 3.2); at least some of them should show the fusant character when the clone really derived from a fusion body. Analysis of the sexual offspring (see Sect. 4) provides further information: more or less suppressed traits should spring off as it was established by Melchers and Labib (1974), Melchers and Sacristán (1977), Power et al. (1979), Aviv et al. (1980), Schieder (1980 b), Evans et al. (1981, 1983), Evola et al. (1983) and Marton et al. (1983). The expected values, however, may be altered by somaclonal variation (Uchimiya et al. 1984; Marton et al. 1985).

Variegated sexual daughter individuals indicate the mixed cell nature of a fusant cell line (e.g., in *Nicotiana:* Nakata and Oshima 1982; Gleba and Sytnik 1984 describing earlier publications of Gleba and coworkers). It must be excluded that biparental transmission of cell organelles leads to de novo formation of mixed cell zygotes.

The reliability of mono- or bifactorial tracer systems is, furthermore, limited by the possibility that new mutants arise in one of the parental lines, simulating the action of the corresponding allele or gene of the other parent. The application of double mutants helps to avoid this incertainty (cf. Evola et al. 1983). Nonallelic secondary mutants can also be excluded by investigating sexual progenies, which, however, is only possible with fertile hybrids.

Loss of marker genes during the development of fusants must be taken into account when devising selection procedures – even in so-called stable hybrids (see Sect. 4.1). Therefore, it is advisable to apply selection as early as possible if it is intended to catch every fusant clone. The later the selection is done, the higher is the probability that clonal variants are selected for which may have lost genetic traits except the characters used in selection. This may be taken as an advantage in cases in which a trait utilized in selection is identical with or closely linked to a property which should be introduced into a genotype, for instance a resistance into a crop.

Most reliable transmission of all genetic entities or a particular nonselective trait to the plant stage is achieved following another procedure: it is then useful to preferentially induce early organogenesis; selection and identification of hybrids may be delayed to a later time (cf. Binding et al. 1982 a). This is especially indicated if fast regeneration can be obtained, for instance by applying the high density streak plating technique (Binding and Kollmann 1985).

The above-mentioned limitations in identification of fusants are negligible in interspecific hybrids, which usually show up a number of hybrid characters supporting the conclusion of their hybrid nature (see for instance Gleba and Hoffmann 1978; Melchers et al. 1978; Binding et al. 1982a). Interspecific hybrids with uniparental nuclei associated with heterospecific organelles (cybrids) are preferentially identified by DNA fragment probing (see Sect. 2.8).

3.2 Separation of Variants

The separation of monotypic cell lines from somaclonal variant mosaics is obtained most efficiently by single cell cloning which is best achieved by regeneration from isolated protoplasts (Binding et al. 1982a; Shepard et al. 1983). The technique is – nonessentially – limited by spontaneous fusion (Chap. II, 3.3) probably leading to secondary hybrids of cells derived from different variant cell lines. Such an event, however, has never been observed, for instance, in experiments with periclinal chimeras of *Petunia hybrida* and of *Solanum nigrum* (Binding et al. 1982b). The most reliable pathway of cloning, namely via adventitious embryos, has so far not been established in somatic hybrids. Structures somehow resembling embryos were observed in a cell line of *Atropa* (×) *Datura* (Krumbiegel 1980). Certain success in clonal segregation can be obtained by adventitious shoot formation; this is applied routinely as it is the most significant step in plant regeneration from fusant callus and has also been intensified by organ explantation (e.g., Maliga et al. 1978; Binding et al. 1982a; Evans et al. 1982). The technique is restricted by the origin of shoot primordia from more than one cell and is therefore indicated only in hybrid clones in which protoplast regeneration failed.

A further means for the isolation of somaclonal variants – especially with reference to plastids and mitochondria in matroclinal species – is the production of sexual progenies: Different types of reduced male fertility were found in F1 of fusants of a male fertile parent and a cytoplasmic male sterile one in *Nicotiana* (Aviv and Galun 1980). The sexual progenies of variegated plants of *Nicotiana* (Gleba and Sytnik 1984), for instance, were composed of variegated as well as of white and green seedlings.

The separation of variants via single cells is obviously no guarantee for their homogeneity as new somaclonal variation may arise during proliferation (see Sects. 2.6 and 5.2.2).

4 Regeneration of Somatic Cell Fusants

The developmental results of fusion products can be taken from Table 1 and Table 1 of Chap. II. Most of the intraspecific and several of the intrageneric fusion bodies could be regenerated to flowering plants, as far as monotypic protoplasts of at least one of the parents were susceptible. Moreover, regeneration has even been obtained in some cases in which the parental protoplasts were not capable of organized growth (see Sect. 8.4).

Table 1. Somatic intergeneric hybrids which regenerated shoots or even plants

Daucus carota (×) *Aegopodium podagraria*	Dudits et al. (1979)
Daucus carota (×) *Petroselinum hortense*	Dudits et al. (1980)
Arabidopsis thaliana (×) *Brassica campestris*	Gleba and Hoffmann (1978)
Brassica campestris (×) *Raphanus sativus*	Pelletier et al. (1983)
Datura innoxia (×) *Atropa belladonna*	Krumbiegel and Schieder (1979)
Nicotiana chinensis (×) *Atropa belladonna*	Gleba et al. (1982)
Nicotiana tabacum (×) *Hyoscyamus muticus*	Jia et al. (1983)
Nicotiana tabacum (×) *Salpiglossis sinuata*	Nagao (1983)
Petunia hybrida (×) *Atropa belladonna*	Gosch and Reinert (1978)[a]
Petunia hybrida (×) *Lycopersicon esculentum*	Tabaeizadeh et al. (1983)
Solanum tuberosum (×) *Lycopersicon esculentum*	Melchers et al. (1978)
	Shepard et al. (1983)
Solanum tuberosum (×) *Nicotiana tabacum*	Skarzhynskaya et al. (1982)

[a] Assumed but not confirmed.

A number of intrageneric fusants passed through the various steps of development about as fast as uniparental clones; but also retarded growth was observed (see Sect. 8.5.2.1). Increased proliferation was described, as well (see Sects. 2.1.1 and 8.4).

Organogenesis was significantly retarded in the heterogeneric combinations listed in Table 1 except the potato (×) tomato hybrids. Usually, the development was characterized by gradual decrease of malformations in subclones (see Sects. 5.2.1 and 8.5.2.2). Organogenesis failed in other intra- and intergeneric and in all interfamiliar fusants. Most of these exhibited limited or even no proliferation.

Plant regeneration suffered from insufficient root formation of regenerated shoots in some interspecific and particularly in intergeneric hybrids (see Sect. 8.5.2.2). The restriction was limited to certain clones only, in some of the combinations. It could be overcome by grafting in *Nicotiana glauca* (×) *langsdorffii* (Smith et al. 1976; Chupeau et al. 1978), *Solanum tuberosum* (×) *Lycopersicon esculentum* (Melchers et al. 1978), and *Datura sanguinea* (×) *innoxia* (Schieder 1980a).

Fertility was mainly limited to combinations in which sexual hybrids were also fertile. It was established in the genera *Brassica* (Schenck and Röbbelen 1982; Schenk 1983), *Datura* (Schieder 1977), *Medicago* (Téoulé 1983a, b), *Nicotiana* (e.g., Melchers and Labib 1974; Smith et al. 1976; Aviv et al. 1980; Evans et al. 1980; Douglas et al. 1981b; Uchimiya et al. 1984; Marton et al. 1985), *Petunia* (e.g., Power et al. 1976, 1977, 1978, 1979; Cocking 1977; Bergounioux-Bunisset and Perennes 1980; Izhar and Tabib 1980).

Fusion hybrids of sexually incompatible combinations have been found to be fertile in *Datura innoxia* (×) *discolor* and *D. innoxia* (×) *stramonium* (Schieder 1978), *D. innoxia* (×) *candida* (Schieder 1980a), *Nicotiana tabacum* (×) *nesophila* and *N. tabacum* (×) *stocktonii* (Evans et al. 1981, 1982), *N. tabacum* (×) *repanda* (Nagao et al. 1982), and *Petunia parodii* (×) *parviflora* (Power et al. 1980). Fertility enabling back-crossing to one of the parents was obtained in *Nicotiana tabacum* (×) *glauca* by Uchimiya (1982). Fertility can also be expected in fusants of sexually incompatible species when no hybrid nuclei are formed. This was the case in the clone P80-45-38s of *Solanum nigrum* (×) *tuberosum* containing the *S. nigrum* nucleus and potato plastids (Binding et al. 1982a). Furthermore, unilateral loss of chromosomes may lead to fertile sublines. This is indicated by partial fertility of *Hyoscyamus* (×) *Nicotiana* plants (Potrykus et al. 1984).

No or reduced fertility corresponded either to inability to form flowers (e.g., in *Daucus* (×) *Aegopodium:* Dudits et al. 1979), or to inappropriate differentiation of the sexual organs (e.g., in *Nicotiana:* Maliga et al. 1978; in *Arabidopsis* (×) *Brassica:* Gleba and Hoffmann 1979), or to failure of the production of functional organs (e.g., in potato (×) tomato: Melchers et al. 1978; Shepard et al. 1983; *Petunia:* Power et al. 1980; in *Solanum:* Binding et al. 1982a; Barsby et al. 1984; in *Atropa* (×) *Datura:* Krumbiegel and Schieder 1981). The particular phenomenon of cytoplasmic male sterility is treated in Sections 7 and 8.5.2.5.

Developmental characteristics must not reflect the properties of the integer fusants of a certain combination. Great diversities may appear between different clones by different fusion events (see Chap. II, 4.2.2.1) or somaclonal variation in the parents or in the fusant clones (see Sects. 5.2. 6, and 7). For illustration, some unpublished data are given concerning *Solanum nigrum* (×) *tuberosum* fusants (Binding et al. 1982a; Gressel et al. 1984): Shoots of 15 clones were strong-growing and easily rooted, but shoots from one clone did not produce any root; regenerants of two fusants contained the nucleus of only *S. nigrum,* plastids of either parent and were fertile, whereas the hybrids were sterile.

5 Fate of Nuclear Traits

This chapter is devoted to observations and investigations on the fates of chromosomes and nucleus-inherited characters in the development of somatic cell fusion hybrids. Usually, two complements of chromosomes are comprised in hybrid nuclei. In the sexual cycle, the complete sets of chromosomes stay together up to the meiosis. Only a few cases have been found in which somatic chromosomal instability leads to the reduction of the chromosome numbers during vegetative development (see Sect. 8.5.1). Somehow increased instability was frequently found in unorganized growing cell cultures. Somatic cell hybrids, however, expressed significant high degrees of somaclonal variation in a number of interspecific combinations. This seems to parallel the corresponding findings in mammalian cell hybrids (Willecke 1978). Discrimination between irregularities caused by culture conditions and by hybrid nature is difficult (see Sect. 8.5.2.3).

5.1 Stable Hybrids

Intraspecific and several intrageneric hybrids expressed high stability with respect to nuclear markers. This was expected from the relationship of the compiled genomes which differed only in one or a few alleles. Chromosomal stability is best documented by fertility of the hybrid plants (Sect. 4). However, variation in chromosome numbers has also frequently been observed in such relatively stable hybrids (Sect. 5.2). Melchers and Labib (1974) already attributed this phenomenon to the numbers of fused protoplasts on the one hand, and to the rather common instability of cell cultures on the other. Shortening of the unorganized growth phase by appropriate culture conditions is therefore demanded in order to obtain increased cytogenetic stability in hybrids of related genotypes.

5.2 Clonal Variation

Clonal variation is illustrated in the following sections with respect to phenotypic variability within clones and between different cell lines which derived by fusion of protoplasts of the same parents, and with respect to chromosomal instability. Possible correlations between phenotypic and chromosomal variability will be analyzed. Control mechanisms are so far not known. Some ideas will be discussed in Section 8.5.2.

5.2.1 Phenotypic Properties

Developmental irregularities of variable expression have been found frequently in interspecific fusants. In closely related species, this phenomenon was often restricted to a few clones, whereas other cell lines developed normally. Growth and differentiation were retarded in clones of hybrid *Datura* (Schieder 1977), *Petunia* (Cocking et al. 1977; Power et al. 1980), as well as in the intergeneric fusants listed in Table 1.

Whereas developmental irregularities indicated different degrees of interferences in the hybrids, a number of characters appeared in the form of clonal and somaclonal variants which reflected significant parental or hybrid characters. Segregation of particular markers appeared as spots and double spots, for instance, in hybrids containing the Su mutant of tobacco (Gleba and coworkers, see Gleba and Sytnik 1984; Evans et al. 1980, 1981, 1983), or in chimeric organization, as for instance in *Solanum* (×) *Lycopersicon* (Poulsen et al. 1980), in *Datura* (×) *Atropa* (Krumbiegel and Schieder 1981), and in *Arabidopsis* (×) *Brassica* (Hoffmann and Adachi 1981).

A number of further properties varied during the development of hybrids: Disappearance of protein bands has been reported for the intergeneric fusants *Glycine max* (×) *Nicotiana tabacum* (Wetter 1977), *Arabidopsis thaliana* (×) *Brassica campestris* (Gleba and Hoffmann 1979), and *Physalis minima* (×) *Datura innoxia* (Gupta et al. 1984). High variability in the shape of leaves was found in *Arabidopsis* (×) *Brassica*, in *Nicotiana* hybrids (Glimelius and Bonnett 1981; Glimelius et al. 1982), and in *Lycopersicon* (×) *Petunia* (Tabaeizadeh et al. 1983). Considerable diversity of intermediate and peculiar appearance (for instance in *Solanum nigrum* (×) *tuberosum* clone P80-45-13: Binding et al. 1982a and unpublished) indicates multiple genetic control and gene dosis effects in leaf morphology (Sect. 8.3). Differences in hair morphology (in *Solanum nigrum* (×) *tuberosum* clones and subclones) and in the ability to form tubers (in the *Solanum* clone P80-45-13) appeared, furthermore, in the vegetative development. Sizes and shapes of flowers were modified, for instance, in *Arabidopsis* (×) *Brassica, Nico-*

tiana tabacum (×) *nesophila* (Evans et al. 1982) and *Solanum nigrum* (×) *tuberosum;* flower pigmentation varied in *Solanum tuberosum* (×) *Lycopersicon esculentum* (Melchers et al. 1978), as well as in the *N. tabacum* (×) *nesophila* hybrids. Most interestingly, fertility was established in some flowers of the intergeneric hybrid *Hyoscyamus* (×) *Nicotiana*.

5.2.2 Fates of Chromosomes

The chromosomes of heterokaryotic fusant cells are combined into a hybrid nucleus preferentially during synchronized mitosis (see Chap. II, 4.2.2.3). They were occasionally found to be arranged in uniparental groups even after a number of cell generations in *Glycine max* (×) *Nicotiana glauca* (Kao 1977), *Glycine max* (×) *Vicia hajastana* (Constabel et al. 1977), *Vicia faba* (×) *Petunia hybrida* (Binding and Nehls 1980) and *Atropa belladonna* (×) *Nicotiana chinensis* (Gleba et al. 1983).

Polyploidy, aneuploidy, chromosome mutations and mitotic irregularities were frequently found in somatic hybrids. It has already been mentioned that polyploidy may be caused by particular fusion events (Chap. II, 4.2.2.1) and that aneuploidy might have been derived from a parent, especially in the case of a cell culture (Sect. 4). However, clonal variation within a fusant must be a consequence of events during proliferation after fusion.

A main reason for the heterogeneity in chromosome number within, but also between somatic hybrid clones seems to be loss of chromosomes which occurred not only in remote combinations (see Table 1), but was also discussed even for intraspecific fusants (*Datura innoxia:* Schieder 1977; *Nicotiana tabacum:* Melchers and Sacristán 1977). Occasionally, drastic reduction down to a few chromosomes from one species was noticed in *Vicia faba* (×) *Petunia hybrida* (Binding and Nehls 1978), in *Datura innoxia* (×) *Atropa belladonna* (Krumbiegel and Schieder 1981), and in *Daucus carota* (×) *Petroselinum hortense* (Dudits et al. 1980). In the case of *Vicia* (×) *Petunia,* alternately unilateral loss of chromosomes from one or the other parent was found in different clones (see Sect. 8.5.2.3).

Several irregularities have been observed in mitosis, especially in intergeneric fusants listed in Table 1 and *Glycine max* (×) *Nicotiana glauca* (Kao 1977), *G. max* (×) *N. tabacum* (Chien et al. 1982), *Vicia faba* (×) *Petunia hybrida* (Binding and Nehls 1978). The described irregularities were a sticking together of chromosomes, chain formation, and multiconstrictional chromosomes, ring chromosomes and anaphase bridges, as well as unusual sizes and heteropycnosis of chromosomes.

Meiosis has been investigated in only a few cases. It is self-evident that multivalent formation was frequent as in polyploids (Scowcroft and Larkin 1981). The disturbed location of chromosomes in meiotic cells of potato (×) tomato hybrids indicated loss of these chromosomes (Shepard et al. 1983).

5.2.3 Correlation of Phenotypic and Cytogenetic Events

Phenotypic variation of nucleus-coded characters may be to some extent caused by epigenetic processes. However, in the majority of cases, especially when the alterations are stable, it can be supposed that somatic mutations are responsible for somaclonal variation of Mendelian properties. This is in agreement with the observation that high morphological variability was associated with chromosomal instability (in the intergeneric hybrids listed in Table 1, as well as in the *Solanum nigrum* (×) *tuberosum* clone P80-45-13: Binding et al. 1982a and unpublished). Furthermore, hyperploidy (Melchers et al. 1978) and chromosome elimination (Poulsen et al. 1980; Shepard et al. 1983) were attributed to morphological heterogeneity in potato (×) tomato.

Correlation of a particular character to a certain cytogenetic process can be supposed with respect to a number of experimental results: retarded organogenesis and gradual decrease of malformations in heterogeneric fusants is most likely due to the formation of asymmetric hybrids by the loss of chromosomes of one parent. The reappearance of traits which were suppressed in the complete hybrid can be taken as indication of the loss of a chromosome which carried the suppressing gene (Schieder 1980; Hoffmann and Adachi 1981; Krumbiegel and Schieder 1981). Loss and additional rearrangement of chromosomes has been taken to explain the presence of *Petunia* isoenzyme bands devoid of detectable *Petunia* chromosomes in a *Parthenocissus tricuspidata* (×) *Petunia hybrida* cell line (Power et al. 1975), and the presence of DNA hybridizing to tobacco-specific DNA in *Hyoscyamus muticus* (×) *Nicotiana tabacum,* which organized shoots resembling *Hyoscyamus* (Jia et al. 1983; Saul and Potrykus 1983, 1984; Potrykus et al. 1984). Evidence for somatic crossing-over was obtained by Evans et al. (1980, 1981, 1983): double spots appeared on leaves of interspecific somatic hybrids of *Nicotiana* with the homozygotic sulfur mutant (Su/Su) as one parent. The participation of transposable elements in the high frequency spot formation (superspot) of *N. tabacum*-Su/Su (×) *sylvestris* was taken into consideration (Evans et al. 1983).

Marton et al. (1985) discussed variable segregation ratios of parental traits (NR^- and albino) in F1 and F2 of somatic hybrids of *Nicotiana plumbaginifolia* to be caused by chromosome abnormalities. Different intensities of the phosphoglucomutase banding patterns in the hybrid progeny of *Nicotiana debneyi* were supposed to be a consequence of meiotic quadrivalent formation (Scowcroft and Larkin 1981).

5.3 Transfer of Small Genome Fractions

It was illustrated in the preceding sections (5.2.2, 5.2.3) that unilateral hybrids containing markedly reduced chromosomal material of one of the parents appeared in somatic hybrids especially in remote combinations. Probably the most advanced example of such development was found in the *Petunia* (×) *Parthenocissus* hybridization experiment (Power et al. 1975): a callus line growing under conditions which did not allow proliferation of uniparental cells contained a nuclear coded enzyme fraction of *Petunia,* but no *Petunia* chromosomes were visible. Either gradual loss of *Petunia* chromosomes or incorporation of a single *Petunia* chromosome into a *Parthenocossus* telophase during a synchronized mitosis of the heterokaryon might have supposingly preceded a translocation.

Artificial promotion of unilateral loss of chromosomes was obtained – for the first time – by Dudits et al. (1980): Fusion of carrot protoplasts with X-ray-inactivated parsley protoplasts resulted in a cell line which was capable of organizing shoots, whereas the entire hybrids were not susceptible. Isoenzyme patterns proved that the parsley chromosome contingent was reduced. In a similar approach, Gupta and coworkers (Gupta et al. 1982; Gupta and Schieder 1983) used entire protoplasts of the nitrate reductase-deficient tobacco line cnx68 as "receptors" and X-irradiated protoplasts of wild-type lines of *Datura innoxia* and *Physalis minima* as "donors". Cell lines could be isolated which revealed only a few out of diverse enzyme bands of the respective donor parent; the corrected character – the reductase activity – was lower than in the respective donor species. Analogous results were obtained with the correction of chlorophyll deficiency of *Datura innoxia* by fusion with treated *Physalis minima* protoplasts (Gupta et al. 1984). Two lines were analyzed, one of which showed up a tetraploid *Datura* set and three chromosomes of *Physalis*, the other was octoploid with one *Physalis* chromosome. The phenomenon that the corrected traits were less pronounced in the hybrid than in wild-type lines of the parents may be explained by minor gene expression or simply by a gene dosis effect.

Genetic correction by the use of X-irradiated protoplasts in somatic hybridization has been discussed as an approach to transformation to some extent resembling gene transfer by more sophisticated techniques of gene technology. The most important aspect of unilateral hybrids, however, is the feasibility of acquiring fertility which makes possible back-crossing breeding (see Sect. 5.2.3).

6 Fate of Plastids

The heteroplasmic state of a cell which is generated by the fusion of biparental protoplasts is an almost unique feature disregarding the few exceptions of organelle transfer in zygote formation. The behavior of cell organelles in mixed cytoplasms of an altered genetic environment has been the objective of many investigations.

Following up the fate of plastids, similar as of mitochondria, a number of questions arose. These include: is the heteroplasmic state stable or is it a transient situation which will end up in the sorting-out of one or the other parental plastid type? If sorting-out occurs, is it random or is it subjected to a principle of any kind? Is sorting-out complete and independent of mitochondria? Might recombination of plastids of different origin be expected and under which conditions could it be detected?

So far, none of these questions has been answered with general validity but a considerable amount of knowledge has been accumulated.

Birkey (1978) regarded segregation of cytoplasmic traits as a typical characteristic, thus the heteroplastic state is earlier or later followed by the homoplastic one. This was confirmed by many authors, who registered alternate loss of one or the other plastid marker in different fusant clones (Belliard et al. 1978; Melchers et al. 1978; Aviv et al. 1980; Iwai et al. 1980; Poulsen et al. 1980; Douglas et al. 1981; Scowcroft and Larkin 1981; Binding et al. 1982a; Uchimiya 1982; Müller-Gensert et al. 1983; Pelletier et al. 1983; Gressel et al. 1984). In all of these cases the segragation of plastids was found to be complete, at least within the limitations of the detection method.

According to Akada et al. (1983), who analyzed plastid segregation at a comparably early callus stage, it is unlikely for complete sorting-out to be a slow process. It seems rather to happen at the very first cell divisions following the fusion. Nevertheless, prolonged expression of traits of either parental origin could be demonstrated. Gleba et al. (1979) analyzed plants which exhibited chlorophyll deficiency as marker of one parent and the large subunit pattern of the ribulose-1,5-bisphosphate carboxylase of the other. The presence of large subunits of both origins was monitored in several regenerants months after protoplast fusion (Chen et al. 1977; Glimelius et al. 1981; Iwai et al. 1980, 1981; Akada and Hirai 1983).

Unless single cell progeny are tested, misleading results from the formation of chimeric plants cannot be ruled out completely. Protoclones were analyzed by Binding et al. 1982a, F1 plants by Fluhr et al. (1983); see also Section 3.1.

The data from literature suggest that mixed populations of plastids are only seldom conserved through numerous cell generations. Indeed, segregation seems to be the common fate. Reasons for this phenomenon have often been discussed. Assuming that there is no obvious disadvantage for plastid replication due to selective pressure or incompatibility with the nuclear genome, and assuming the numbers of either plastid types to be about the same, sorting-out should be random (Scowcroft and Larkin 1981; Galun et al. 1982b). On the other hand, there are results which support evidence for a nonrandom unidirectional segregation of plastids (Flick and Evans 1982; Kumar et al. 1982). Insufficient genome-plastome interaction, if not complete incompatibility, must be taken into account, which results in underreplication of one of the organelle types.

If plastids could be regarded as behaving like populations rather than as individuals, this might further contribute to an understanding of the speed and completeness of the segregation process. Physiological conditions of a newly formed hybrid cell, e.g., lack of synchronization of biochemical pathways, may also be disadvantageous to only one plastid type despite complete nuclear-plastidic compatibility.

Since the number of samples so far analyzed is still small and more basic knowledge must be accumulated, the interpretation of phenomena remains necessarily speculative.

The mechanisms involved in random or nonrandom segregation of plastids may be discussed also when plastidic and mitochondrial traits are transmitted. Results given by some authors who found independent segregation of plastidic and mitochondrial traits suggest that the underlying principle might be random (Belliard et al. 1978; Bonnett and Glimelius 1983; Fluhr et al. 1983). Other experiments, however, which resulted in the preferential transmission of streptomycin resistance (plastidic marker), together with the expression of cytoplasmic male sterility (mitochondrial marker), support evidence that cytoplasmic characters are not completely independent (Menczel et al. 1983).

Recombination of plastids of different origin is an event which has been long sought. It was known only for chloroplast DNA of *Chlamydomonas* (Lemieux et al. 1981), although experiments had been designed to detect even minor rearrangements of the DNA (Schiller et al. 1982; Fluhr et al. 1984). Instead, cosegregation of uniparental plastidic traits, such as large subunit pattern of the ribulose-1,5-bisphosphate carboxylase, chlorophyll deficiency and antibiotic resistance was observed (Dix et al. 1977; Medgyesy et al. 1980, 1983; Cseplö et al. 1983; Gressel et al. 1984; Chetrit et al. 1985).

Plastidic DNA can be described as a molecule of three recombinational units, a large and a small unique region and the inverted repeats. Fluhr et al. (1983) stressed the probability of recombinational processes between these regions. Most recently, Medgyesy et al. (1985) could in fact demonstrate a replacement in the large single-copy unit of *Nicotiana plumbaginifolia* by the homologous region of *Nicotiana tabacum*. This is the first direct observation of ptDNA recombination in higher plants. The fact that it was found only under strong double-selective pressure of two different antibiotics might be an indication of the paucity and perhaps of the instability of such processes.

Segregation of different plastid types, or even recombination between them, could be one of the nearest applications of somatic cell genetics to plant breeding. Plastids code, for instance, for herbicide resistance and can be matched by fusion to a nucleus of a cross-incompatible species. In experiments with a triazine-resistant *Solanum nigrum* and *S. tuberosum*, a combination of the *S. nigrum* nucleus with *S. tuberosum* plastids was obtained (Binding et al. 1982a; Gressel et al. 1984). Pelletier et al. (1983) recovered rapeseed plants from triazine-resistant *Brassica campestris*. Similar attempts were made by Robertson and Earle (1985), combining broccoli with an atrazine-resistant *Brassica campestris* line; in this case, among 1300 protoplast-derived calluses, one survived in a medium containing atrazine and could be regenerated to a plant which exhibited mainly characteristics of broccoli but harbored atrazine-resistant chloroplasts.

7 Fate of Mitochondria

As discussed in the previous Section, cell organelles are located in a novel and maybe unstable situation after the combination of two different cytoplasms. Mitochondria should therefore be submitted to the same hypothetical mechanisms of selective disadvantage, underreplication, sorting-out, and recombination as their plastidic counterparts. In general, this assumption proves right. There are some differences, however, which become evident by the frequency of phenomena as compared to those occurring to plastids.

The coexistence of mitochondrial traits from both parents for numerous cell generations has also only rarely been observed (e.g., Izhar et al. 1983). The true biparental transmission to the mitotic progeny and thus the exclusion of chimera formation could in this case be confirmed by the segregation of fertility and cytoplasmic male sterility in the sexual F2 progenies.

Complete sorting-out of one of the parental mitochondrial traits was reported by Izhar and Tabib (1980). From a single fusion experiment they recovered cms as well as fertile plants. The latter were self-fertile and gave rise to either stable male sterile or fertile lines. Since this segregation happened from a single phenotypically fertile cytoplasm, it is obvious that the heteroplasmic state had been carried on until meiosis.

Co-segregation of cytoplasmic male sterility and flower malformation, together with the finding of an almost unchanged restriction pattern of the "fertile" mitochondrial DNA, was also regarded as unilateral sorting-out of parental mitochondria (Aviv et al. 1984). Reviewing reports on the fate of mitochondria in mixed cytoplasms, both carrying on the heteroplasmic state as well as complete sorting-out, seem to be exceptional characters. In most cases, altered patterns of mtDNA restriction fragments were found. In one of the first investigations (Belliard et al. 1979), new bands occurred after restriction cleavage in a tobacco hybrid. This was interpreted as the formation of a recombinant molecule of either parent's origin.

Much information on the mitochondrial DNA architecture has been accumulated since then, and somatic hybridization experiments have contributed much to stimulate this research. The mitochondrial genome is composed of multiple circular DNA's of different sizes which are not necessarily subgenomic copies of a master molecule (Boeshore et al. 1983). Additional plasmid-like circles may be present to increase the complexity and plasticity of the mitochondrial genome and obscure understanding it. For this reason, several authors who registered new restriction bands or intermediate patterns discussed their results with respect to true intermolecular recombination as well as to possible intramolecular rearrangements of parts of an uniparental genome (Nagy et al. 1981; Galun et al. 1982; Boeshore et al. 1983). Even hybridization of parental sequences to mitochondrial DNA of fused cytoplasms did not give unequivocal proof for either intermolecular recombination or intramolecular rearrangement (Nagy et al. 1981).

Rearrangement of mitochondrial DNA as a result of prolonged tissue culture is further making the interpretation of such findings in somatic hybrids difficult. Gengenbach et al. (1981) regenerated fertile maize plants from cultures of "Texas" cytoplasmic male sterile cells. Creation of a heteroplasmic state by protoplast fusion, however, was a necessary prerequisite for the detection of changes in the mtDNA restriction patterns (Nagy et al. 1983). Homoplastic fusion was not sufficient to generate new bands, thus triggering effects of the fusion process itself can be ruled out.

In a recent work including *Brassica napus* and *Raphanus sativus* cytoplasms, Chetrit et al. (1985) strongly support evidence for true mitochondrial recombination. Their findings of DNA polymorphism are in accordance with what could be expected from physical maps of the *Brassica* parent. As an additional item, an insight into coding for cytoplasmic male sterility was made possible. The sequences responsible for the expression of cms were found to be localized on one of the genomic molecules rather than on a plasmid. There was not even a correlation between the degree of cms expression and the amount of plasmid DNA present in a cybrid cell line. Further development and application of mtDNA hybridization probes will elucidate much of the present uncertainties concerned with the fates of mitochondria after somatic hybridization, even though it is unequivocally evident that coexistence or complete segregation of both of the parental mitochondria types without modification of their genomes is the exception. Rearrangement to minor or extensive scale seems to be the common feature if cytoplasms of different origins are combined. This is strikingly different from the behavior of plastids (see Chap. 6) under the same conditions and gives an indication of the enormous plasticity of the mitochondrial genome.

Transmission of single traits such as cytoplasmic male sterility by mitochondrial recombination to an otherwise unchanged genetic background from nucleus and plastids may be helpful in some of the problems which the plant breeder has to face.

8 Interspecific Interactions

8.1 Introduction

Experimental protoplast fusion opens up the fascinating possibility of overcoming even the strongest sexual incompatibilities and hence combining genotypes which had been separated by sexual barriers for more or less long periods of evolution. The feature has been repeatedly discussed especially with respect to experimental protoplast fusion, recently by Harms (1983).

The application of fusigenic agents makes possible fusion of protoplasts of any taxonomic position. Sexual incompatibilities preventing zygote or embryo formation do not play any role in the very process of fusion as far as it appeared from the performed experiments.

Mutual reactions of the combined systems may become evident in gene expression, cytoplasmic metabolism, organelle transmission, regulation of developmental steps, mitosis, and meiosis. Detailed investigations of the phenomena can be utilized to answer the following questions:

a) How conservative are certain metabolic or developmental processes?
b) To what extent is gene expression altered in the synthetic genetic backgrounds?
c) How far is co-evolution of nuclear genes, plastomes and chondriomes progressed in different species as compared to their degrees of relationships?
d) Are the genome/plastome incompatibilities which are known from sexual investigations (e.g., in *Oenothera:* cf. Stubbe 1964; Stubbe and Raven 1979; and *Epilobium:* cf. Michaelis 1955) established as specialized barriers between closely related species, or do they reflect evolutionary steps in more common diversification of species which have been excluded from sexual recombination by other primary barriers?
f) Are physiological processes of intraspecific sexual incompatibility expressed in experimental protoplast fusion and, consequently, can fusion experiments contribute to the analysis of sexual incompatibility?

g) Which genetic factors do control nucleus segregation and hybrid nucleus formation; chromosome multiplication, reduction, fragmentation and recombination; as well as plastid and mitochondria segregation and their parasexual processes?
h) To what extent is the application of experimental cell fusion genetics to breeding programs limited by heterospecific incompatibilities, or can particular incompatibilities even be exploited in breeding?

Experiments may be devised to obtain answers to the questions raised rather than that the available information can be used for clarification. So far, only preliminary and more or less speculative ideas emerged from investigations on experimental protoplast fusion, grafting and haustoria formation. The numbers of respective experiments in which the development and metabolism, as well as the fates of nuclear and extrakaryotic genophores have been analyzed with appropriate methods and in sufficient detail is rather low.

Increasing attention with respect to interspecific interactions is being paid to grafting and plant/plant parasitism (see Kollmann et al. 1985). As heterospecific plasmodesmata are formed, some phenomena may be expected which correspond to observations with fusants.

8.2 Co-Existence of Heterospecific Genomes and Plasmones

Clonal proliferation of fusants indicated high degrees of compatibility in basic metabolism even in combinations of remote species, provided that both genotypes were expressed (cf. Constabel et al. 1975). Chromosomes of different classes were integrated into metaphase plates; secondary metabolites which were poisonous to the partner organism did not affect proliferation of the fusant (Constabel et al. 1976). A strange phenomenon is the congregation of related chromosomes in intergeneric hybrids (see Sect. 5.2.2). Cooperation of nucleus and foreign plastid genes was demonstrated by hybrid fraction I proteins (see Sect. 2.8.3) and, in whole complexity, by the viability and photosynthetic activity of nucleus-plastid recombinant cybrids of *Nicotiana sylvestris* (×) *tabacum* (Zelcer et al. 1978; Aviv and Galun 1980), *N. sylvestris* (×) *plumbaginifolia* (Czeplö et al. 1983, 1984), and *Solanum nigrum* (×) *tuberosum* (Binding et al. 1982a; Gressel et al. 1984). The *Solanum* cybrid was fertile, in contrast to the nuclear hybrids.

Intracellular coexistence of different types of plastid over long developmental periods was concluded by Gleba et al. (1985) in *Nicotiana* hybrids from the occurrence of biparental plastids in F1. This conclusion is unequivocal when exclusively maternal transmission of the plastids is guaranted.

Sorting-out of chromosomes and plastids cannot simply be taken as indication of limited compatibilities. It must be regarded that mitotic disturbances are a common feature in plant tissue culture (cf. Binding and Nehls 1980); unilateral loss of plastids may be a consequence of random distribution, even though early segragation is not easily explained by this submission (see Sect. 6). The interpretation of the experimental data on the mitochondria is nearly impossible, so far (see Sect. 7).

8.3 Expression of Characters Within the Parental Limits

Manifold types of phenotypic interaction of alleles, genes, chromosomes, and cell organelles are known from sexual crosses. It is more or less evident that no basically new mechanisms could be expected from experimental cell fusion genetics. Some of the interactions of nuclear traits may be controlled by different alleles of one gene. Proof is only possible in related species in which chromosome homologies suffice for meiotic pairing and the sexual offspring can be investigated. It is reasonable in all other cases to use the terminology of intergeneric relationships, as for instance epistasis, hypostasis, and suppression.

A number of parental characters were expressed in a dominating manner in the somatic hybrids. This concerned drug resistance (see Sect. 2.2.2), morphology (see Sect. 2.4), protein patterns (see Sect. 2.8), and low molecular compounds (see Sect. 2.9). In evaluating experimental results with respect to independently expressed characters it is important to consider the possibility that chimeric tissue may have been investigated. The probability of misinterpretation can be reduced by utilizing single cell clones; but new somaclonal variation may appear during the proliferation of a hybrid cell (see Sect. 3.2). Independent expression may also be stimulated by loss of a genetic trait which would have acted as a counterpart to a certain character in the integer hybrid. Characters are more conclusive when they are clearly localized; as, for instance, *Solanum nigrum* (×) *tuberosum* plants formed hairs which were as long as in potato but had a gland cell like *S. nigrum* (Binding et al. 1982a).

Intermediate expression of parental characters was, for example, described by Smillie et al. (1979) concerning the chilling resistance of potato (×) tomato hybrid plants. The judgement of intermediate morphologies of leaves or flowers is again difficult by the possibility that a certain shape may also be a consequence of chimeric nature of the organ (see Sect. 3.1). This could be ruled out, for instance, in subclones of the instable hybrid clone P80-45-13 of *Solanum nigrum* (×) *tuberosum* of which variants in leaf shapes were isolated (Binding et al. 1982a) showing high degrees of phenotypic stability after subcloning via adventitious shoots (Binding, unpublished).

Polygeneic control of morphological characters can be concluded from the appearance of variability in instable hybrids. This has been observed, for instance, in *Solanum nigrum* (×) *tuberosum* where the leaf shapes ranged from *S. nigrum* type in the complete hybrid over different types of denticulation and insection to integer and pinnate leaves somehow resembling leaf types of the potato parent (Binding et al. 1982a and unpublished).

8.4 Hybrid Characters Exceeding the Parental Phenotypes

A few fusion combinations have been described in which the fusants had gained properties which were not expressed equally in a parental line. The restauration of organogenetic activity (Maliga et al. 1977; Gleba and Hoffmann 1979, 1980; Marton et al. 1984; Potrykus et al. 1984) and the independence from external phytohormones in hybrids of *Nicotiana* (Carlson et al. 1972; Smith et al.

1976), *Datura* (Schieder 1980a), and *Hyoscyamus* (×) *Nicotiana* (Potrykus et al. 1984) are some examples. The heterotic increase of cell proliferation which appeared repeatedly (in *Datura:* Schieder 1977, 1980a, 1982; in *Arabidopsis* (×) *Brassica:* Gleba and Hoffmann 1978; in *Atropa* (×) *Datura:* Krumbiegel 1980; in *Nicotiana:* Douglas et al. 1981a) may also be explained by increased internal hormone levels. Tobacco hybrids and cybrids formed giant flowers (Glimelius and Bonnett 1981; Glimelius et al. 1981). A new esterase isoenzyme band in *Arabidopsis* (×) *Brassica* which was not due to hybrid protein association was supposed to be caused by gene derepression in the hybrid (Gleba and Hoffmann 1979).

8.5 Incompatibility and Related Phenomena in Hybridization

High degrees of congruity in basic cell metabolism in spermatophytes was concluded from the development of interfamiliar fusants (see Sect. 8.2). On the other hand, several phenomena in the sexual cycle, in the development of fusants, in grafting and host/parasite interaction have been and partially must be explained by suppression, incompatibility or incongruity.

As the experimental distinction between these phenomena is still unsatisfactory in sexual and somatic processes, the term "incompatibility" will be used here comprehensively for all phenomena indicating restrictions in joint processes in physiology and development as a consequence of the combination of cells by the sexual process, by induced fusion, by grafting and by plant/plant parasitism (cf. also Harms 1983). Morphological incompatibilities, as for instance heterostylism, developmental incompatibility like protandrism and proterogynism, and all types of intercellular incompatibility passing the apoplast – such as pollen/stigma interaction – will not be considered in this volume which is devoted to the development of fusion products. Topic of this section are processes within a cell or a symplast accounting for incompatibility.

8.5.1 Incompatibility in the Sexual Cycle

Manifold types of sexual incompatibility are established in the life cycle of higher plants which help to maintain heterozygosity or to prevent interbreeding. The incompatibilities act in the course of pollination and zygote formation, embryo- and plant development, meiosis, and gametogenesis. Intracytoplasmic incompatibility can begin with the pollen tube-embryo sac fusion. Embryo formation may be impeded not only by intracellular factors, but also by intercellular phenomena, that is suggested when the growth of embryos is restored by in-vitro culture (see Raghavan 1977). Impaired development of hybrid plants, however, is unequivocally based on intracellular incompatibilities. They depend either on genomic imbalance, on the interference of secondary metabolites or on incompatibilities between cell organelles.

The discrimination between these alternatives or even the detection of a particular mechanism has not been possible in most cases. As one exception, some insight could be obtained into alloplasmatic lines exhibiting cytoplasmic male sterility by the interaction of the mitochondria with foreign genetic background and being restored by single nuclear genes (cf. Frankel and Galun 1977; see Sect. 7).

The loss of chromosomes in *Hordeum* interspecific hybrids is another well-investigated example of incompatibility established by sexual crossing: Kasha and Kao (1970) detected unidirectional elimination of the *Hordeum bulbosum* chromosomes in the early development of hybrid embryos. The process finishes up with a haploid set of the other parent, *Hordeum vulgare*. The chromosome elimination is apparently controlled by a complex genetic system. Ho and Kasha (1975) found that both arms of chromosome 2 of *H. vulgare* and the short arm of chromosome 3 were involved. Imbalance of the genomes was suspected by Subrahmanyam and Kasha (1973). Similar events have been found in sexual hybrids *Triticum aestivum* (×) *H. bulbosum* (Barclay 1975), *H. vulgare* (×) *Secale cereale* (Fedak 1977) and in *Nicotiana* hybrids (Gupta and Gupta 1973).

Sexual incompatibility in meiosis has frequently been observed and investigated in interspecific hybrids. It was usually explained by reduced chromosomal homologies. Impaired development of gametophytes and gametes was mostly a consequence of meiotic anomalities.

8.5.2 Incompatibility in Fusants

Several authors used the term incompatibility in induced fusion genetics for all cases in which the fusants exhibited more minor developmental activities than the superior one of the parents. This definition includes cases in which simply further development is suppressed by a factor of the inferior partner which limits equally the growth of its uniparental cultures. Such a situation may also be realized when the growth of one parent was impaired by selective culture conditions. In the following, the term incompatibility is used to describe developmental retardation or impairment, which are significantly more pronounced than in cultures of either parent in the same experiment. This finding must be statistically ensured to avoid misinterpretation by the variability in protoplast regeneration. In particular the last demand was hardly realized in most investigations.

8.5.2.1 Incompatibility in the Formation of Fusant Cell Clones

No limitations have been detected in the formation of fusion bodies which could be reduced to incompatibility (see Chap. II); but some incompatibility reactions were evidenced in early development of fusants (see Chap. II, 4.2).

Callus formation was possible even in interfamiliar combinations [for instance, *Glycine max* (×) *Nicotiana glauca:* Kao 1977; *Vicia faba* (×) *Petunia hybrida:* Binding and Nehls 1978; and – with highly reduced hybrid nature – *Parthenocissus tricuspidata* (×) *Petunia hybrida:* Power et al. 1975]. Retarded growth as it was found, for instance, in *Petunia* (Cocking et al. 1977; Power et al. 1980), in *Datura* (Schieder et al. 1977) and in Apiaceae (Dudits et al. 1979) is supposingly an expression of interference of the combined genotypes. It can be taken from Table 1 in Chap. II that much more interfamiliar combinations revealed no or limited proliferation. Considering the contribution of incompatibility it must be stressed that in most cases of incompetent fusants also at least one of the parents exhibited equal response to the culture conditions. On the other hand, sustained divisions of fusants cannot be taken as prove for compatibility of the combined genotypes. Cell lines may get rid of a sublethal incompatibility by segregation of cell organelles (see Sects. 6 and 7) or loss of chromosomes (see Sect. 5) and may then overgrow the retarded cells. As another probability, suppression of parts of a genome has been repeatedly discussed.

8.5.2.2 Incompatibility in Organogenesis

Criteria of incompatibility acting on the level of shoot formation is: vigorous fusant callus which does or does not restrictively produce organs provided that callus of the partners both showed organized growth readily and to high degrees in the same experiment. This strict situation was apparently met in *Datura innoxia* (×) *Nicotiana tabacum* (Gupta et al. 1982) and in *Hyoscyamus niger* (×) *N. tabacum* (Lázár et al. 1983). Impaired and retarded organogenesis was also found in the other intergeneric hybrids listed in Table 1. Gradually increasing normalization has been discussed in correlation to the loss of chromosomes (see Sect. 5.2.3) which probably lead to a better genetic balance, to the elimination of particular factors of incompatibility, or to a genetic trait which is responsible for the lack of organogenesis in one of the parents.

Disturbed growth cannot per se be taken as an indication of interspecific incompatibility. Somaclonal variation may lead to genetic mosaics giving rise to malformations.

Reduced ability to form roots at regenerated shoots has been observed in some interspecific hybrids (*Nicotiana:* e.g., Smith et al. 1976; Chupeau et al. 1978; *Datura:* Schieder 1980a, 1982; *Daucus:* Kameya et al. 1981; *Solanum:* Binding et al. 1982a and in the intergeneric hybrids listed in Table 1). It may be taken as an indication of some kind of incompatibility.

It has been already repeatedly mentioned that phenomena which can be attributed to incompatibility are found at different degrees in different clones and subclones of the same fusant combination. Consequently, the question if certain species are incompatible with respect to organogenesis cannot be answered by the investigation of one or a few fusant clones, and more experimental data must be accumulated to obtain information if and how far the ability of organized growth depends on the degrees of relationship of the fusion partners.

8.5.2.3 Incompatibility in the Transmission of Chromosomes

Karyotypic instability has been observed in most of the experimental fusion hybrids. The question has been discussed if loss and rearrangement of chromosomes were induced by incompatibility in a manner similar to that found in the sexual *Hordeum* hybrids (Sect. 8.5.1) or if they are comparable to similar events in uniparental tissue. This latter suspicion was suggested in one of the *Vicia* (×) *Petunia* protoclones (Binding and Nehls 1980). Impaired mitotic figures in *Glycine max* (×) *Nicotiana glauca* were so pronounced that they can in fact be ascribed to incompatibility (Kao 1977). Other fusants were not investigated sufficiently to discriminate unequivocally between genomal imbalance and common events in tissue culture. Higher instability in somatic hybrids than in sexual hybrids (Evans et al. 1982) may be explained to some degree by the influence of in-vitro conditions to which only the fusants were submitted.

Chromosome loss leads to unilateral development in hybrids of remote species as far as this could be investigated. This feature is easily explained under the submission that the constitutions of the coupling groups were widely varied during evolution. This not only makes regular meiotic pairing impossible, but also chromosome substitution. Cell lines must hence contain at least a complete haploid set of chromosomes of one partner. Initial random loss of one type of chromosome is supposed to decide on the direction of unilateral development into one or the other direction, respectively. The direction can, hence, probably be influenced by the investigator by using a monohaploid line of the parent to be eliminated. Haploids have been so far used only in order to obtain a reduced degree of polyploidy in the hybrid (Melchers and Labib 1974).

8.5.2.4 Incompatibility with Respect to Plastids and Mitochondria

Indications for incompatibility in the interaction of cell organelles have mainly been obtained in investigations on the early development of fusants (see Chap. II, 4). Nearly no preferential loss of a certain organelle type nor cosegregation could be detected (see Sects. 6, 7). It has been discussed if recombination in mitochondria which was concluded from DNA restriction patterns (see Sect. 7) is a usual process in cell organelles which did not became visible before fusant analyses were performed, or if it is a particular peculiarity of fusants or, finally, if it is mainly restricted to the situation of cytoplasmic male sterility. The last suspicion reflects the fact that respective investigations were limited to fusants containing a cms parent.

8.5.2.5 Incompatibility Expressed in Impaired Fertility

All investigated intergeneric and some of the interspecific nuclear hybrid fusant plants were sterile with the exception of an asymmetric hybrid cell line of *Hyoscyamus* (×) *Nicotiana* (see Sects. 4 and 5.2.1). Sterility is not surprising as it is common also in sexual interspecific hybrids. Incompatibility has been discussed particularly in cases in which reproductive organs were lacking or insufficiently developed (see Sect. 5.2.1). Cytoplasmic male sterility, which is a well-known indication of incompatibility in interspecific cybrids (see Sect. 8.5.1), has been transmitted to fusants (see Sect. 6) but was in no case created de novo in fusants.

A number of hybrid plants showed well-formed flower organs, but were sterile (e.g., in *Solanum* (×) *Lycopersicon:* Melchers et al. 1978; in *Solanum nigrum* (×) *tuberosum:* Binding et al. 1982a). Berries, but no functional seeds, were developed; meiotic figures, however, seemed to be ordered enough to suggest the formation of functional microspores (observations also of Binding et al., unpublished). As a matter of fact, speculation of the action of incompatibility in these and many other interspecific hybrids has, so far, no reasonable basis.

8.6 Interactions in Grafting and Parasitism

Informations on the mechanisms involved in somatic incompatibilities may be drawn from graftings and plant/plant parasite combinations as in both systems cytoplasmic fusion occurs even though it is limited to small plasmodesmata (Dörr 1968; Kollmann and Glockmann 1985).

It is well known from the practice of gardeners that varying success in grafting is obtained depending on the partners. Failure of good connections can be taken as an indication of incompatibility if either parant can easily be autografted. This control, however, has not been done in a number of combinations. Incompatibility in grafting and parasitism has been surveyed in a volume edited by Moore (1983).

Processes are of interest here which prevent or disturb the formation of plasmodesmata between stock and scion, or which act in later developmental stages after cytoplasmic continuity has been established. Lack of plasmodesmata must be verified by exhaustive electron microscopic investigations. Significantly reduced numbers of plasmodesmata were found in *Vicia faba* (×) *Helianthus annuus* (Kollmann and Dörr 1985). Some information may hopefully come also from protoplast grafting experiments.

The extent of interaction between the graft partners was conclusively documented by the discovery of Frankel (1971) that cytoplasmic male sterility was transmitted from the stock to the scion in *Petunia*. This finding was transferred to sugarbeet (Curtis 1960) and alfalfa (Thompson and Axtell 1978). As probable alternatives to grafting of plant organs, investigations on the association of isolated protoplasts and cells plated at locally high densities (Binding 1984; Binding and Kollmann 1985) may contribute to the detection and analysis of somatic incompatibility.

Whereas in grafting experiments interspecific combinations are produced, the fitness of which had not been controlled during evolution, plant/parasite interactions may indeed have been tuned by evolutionary mechanisms. This is indicated by limited host species of several parasitic plants. It may hence be expected that specific incompatibility systems may be discovered. Fusion of isolated protoplasts of parasites and their hosts may help to analyze the phenomena.

9 Conclusions and Prospects

A number of interesting perspectives have been opened during investigations on the development of protoplast fusion products. Most important and emphasizing features are (1) that plants could, in fact, be regenerated from fusion bodies of sexually incompatible species and genera, and (2) that fertile plants were obtained with heterospecific cell organelles even in combinations in which the complete hybrids were not able to produce sexual progenies.

The attempt has been made to correlate stability and different degrees of taxonomic relationship of the parents. Whereas some indication of incompatibility and related interactions has been observed, much more investigation is needed to obtain an overall view of mechanisms which reduce regenerative capacities in somatic hybrids. The technologies of protoplast fusion, regeneration of fusant plants, selection of fusant clones and variated subclones, and identification of genetic traits are so far advanced – and still developing – that increasing insight into the fates of chromosomes, plastids, and mitochondria, and in the processes of interaction between entities from species of diverse taxonomic relationships can be expected in the near future.

Protoplast fusion genetics is hence on the way to contribute to the elucidation of incompatibility (in the narrow sense), of conservativity and convergency, as well as of evolutionary variability and divergency of physiological and morphogenetic processes in embryophytes.

Acknowledgment. The authors wish to thank Volker Schroeren for his collaboration in the preparation of the manuscript.

References to Chapter II and III

Ahkong QF, Fisher D, Tampion W, Lucy JA (1975) Mechanisms of cell fusion. Nature (Lond) 253:194–195

Akada S, Hirai A (1983) Studies on the mode of separation of chloroplast genomes in parasexual hybrid calli. II. Heterogeneous distribution of two kinds of chloroplast genomes in hybrid callus. Plant Sci Lett 32:95–100

Archer EK, Landgren CR, Bonnet HT (1982) Cytoplast formation and enrichment from mesophyll tissue of *Nicotiana* spp. Plant Sci Lett 25:175–185

Aviv D, Galun E (1980) Restoration of fertility in cytoplasmic male sterile (CMS) *Nicotiana sylvestris* by fusion with X-irradiated *N. tabacum* protoplasts. Theor Appl Genet 58:121–127

Aviv D, Fluhr R, Edelman M, Galun E (1980) Progeny analysis of the interspecific somatic hybrids: *Nicotiana tabacum* (cms) (×) *Nicotiana sylvestris* with respect to nuclear and chloroplast markers. Theor Appl Genet 56:145–150

Aviv A, Arzee-Gonen P, Bleichman S, Galun E (1984a) Novel alloplasmic *Nicotiana* plants by "donor-recipient" protoplast fusion: Cybrids having *N. tabacum* or *N. sylvestris* nuclear genomes and either or both plastomes and chondriomes from alien species. Mol Gen Genet 196:244–253

Aviv A, Bleichman S, Arzee-Gonen P, Galun E (1984b) Intersectional cytoplasmic hybrids in *Nicotiana*. Identification of plastomes and chondriomes of *N. sylvestris* (×) *N. rustica* cybrids having *N. sylvestris* nuclear genomes. Theor Appl Genet 67:499–504

Barcley IR (1975) High frequencies of haploid production in wheat (*Triticum aestivum*) by chromosome elimination. Nature (Lond) 256:410–411

Barsby TL, Shepard JF, Kembe RJ, Wong R (1984) Somatic hybridization in the genus *Solanum*: *S. tuberosum* and *S. brevidens*. Plant Cell Rep 3:165–167

Belliard G, Pelletier G, Vedel F, Quetier (1978) Morphological characteristics and chloroplast DNA distribution in different cytoplasmic parasexual hybrids of *Nicotiana tabacum*. Mol Gen Genet 165:231–237

Belliard G, Vedel F, Pelletier G (1979) Mitochondrial recombination in cytoplasmic hybrids of *Nicotiana tabacum* by protoplast fusion. Nature (Lond) 281:401–403

Bergounioux-Bunisset C, Perennes C (1980) Transfer de facteurs cytoplasmiques de la fertilité male entre 2 lignées de *Petunia hybrida* par fusion de protoplastes. Plant Sci Lett 19:143–149

Berry SF (1983) Factors affecting somatic hybridization between sexually incompatible species of *Petunia*. In: Potrykus I, Harms CT, Hinnen A, Hütter R, King PJ, Shillito RD (eds) Protoplasts 1983 – poster proceedings. Birkhäuser, Basel, pp 83–84

Binding H (1966) Regeneration und Verschmelzung nackter Laubmoosprotoplasten. Z Pflanzenphysiol 55:305–321

Binding H (1974a) Fusionsversuche mit isolierten Protoplasten von *Petunia hybrida* L. Z Pflanzenphysiol 72:422–426

Binding (1974b) Regeneration von haploiden und diploiden Pflanzen aus Protoplasten von *Petunia hybrida* L. Z. Pflanzenphysiol 74:327–356

Binding H (1974c) Mutation in haploid cell cultures. In: Kasha KJ (ed) Haploids in higher plants. Guelph, Canada, pp 323–337

Binding H (1976) Somatic hybridization experiments in solanaceous species. Mol Gen Genet 144:171–175

Binding H (1979) Subprotoplasts and organelle transplantation. In: Sharp WR, Larsen PO, Paddock FF, Raghavan V (eds) Plant Cell and Tissue Culture. Ohio State Univ Press, Columbus, pp 789–805

Binding H (1984) Aufzucht von Pflanzen aus isolierten Protoplasten und Fusionskörpern. In: Mitteilungsband – Kurzfassungen der Beiträge, Botaniker Tagung in Wien 1984. Inst f Bot, Univ Wien, S 61

Binding H, Kollmann R (1976) The use of subprotoplasts for organelle transplantation. In: Dudits D, Farkas GL, Maliga P (eds) Cell Genetics in higher plants. Akadémiai Kiadó, Budapest, pp 191–206

Binding H, Kollmann R (1985) Regeneration of protoplasts. In: Schäfer-Menuhr A (ed) In vitro techniques. Propagation and long term storage. Proceedings of a seminar in the CEC programme of co-ordination of research on plant productivity, Braunschweig 1984. Nijhoff, Junk, Dordrecht Boston Lancaster, pp 93–99

Binding H, Nehls R (1978) Somatic cell hybridization of *Vicia faba* (×) *Petunia hybrida*. Mol Gen Genet 164:137–143

Binding H, Nehls R (1979) Subprotoplasts and organelle transplantation. In: Sharp WR, Larsen PO, Raghavan V (eds) Plant cell and tissue culture – principles and applications. Ohio State Univ Press, Columbus, pp 789–805

Binding H, Nehls R (1980) Transfer of genetic information in higher plants via protoplast fusion. In: Ferenczy L, Farkas GL (eds) Advances in protoplast research. Pergamon, Oxford, pp 315–319

Binding H, Nehls R (1982) Techniques of somatic cell hybridization by fusion of protoplasts. In: Shay JW (ed) Techniques in somatic cell genetics. Plenum, New York, pp 471–492

Binding H, Weber HJ (1974) The isolation, regeneration and fusion of *Phycomyces* protoplasts. Mol Gen Genet 135:273–276

Binding H, Binding K, Straub J (1970) Selektion in Gewebekulturen mit haploiden Zellen. Naturwissenschaften 57:138–139

Binding H, Nehls R, Kock R (1980) Versuche zur Protoplastenregeneration dikotyler Pflanzen unterschiedlicher systematischer Zugehörigkeit. Ber Dtsch Bot Ges 93:667–671

Binding H, Jain S, Finger J, Mordhorst G, Nehls R, Gressel J (1982a) Somatic hybridization of an atrazine resistant biotype of *Solanum nigrum* and *Solanum tuberosum*. I. Clonal variation in morphology and in atrazine sensitivity. Theor Appl Genet 63:273–277

Binding H, Nehls R, Jörgensen J (1982b) Protoplast regeneration in higher plants. In: Fujiwara A (ed) Plant tissue culture 1982. Maruzen, Tokyo, pp 575–578

Birkey CW (1978) Transmission genetics of mitochondria and chloroplasts. Ann Rev Genet 12:471–512

Boeke JH (1973) The use of light microscopy versus electron microscopy for the location of postgenital fusions in plants. Proc K Ned Akad Wet Ser C Biol Med Sci 76:528–535

Boeshore ML, Lifshitz I, Hanson MR, Izhar S (1983) Novel composition of mitochondrial genomes in *Petunia* somatic hybrids derived from cytoplasmic male sterile and fertile plants. Mol Gen Genet 190:459–467

Boss WF, Mott RL (1980) Effects of divalent cations and polyethylen glycol on the membrane fluidity of protoplasts. Plant Physiol (Bethesda) 60:835–837

Boss WF, Allen NS, Grimes HD (1983) Developmentally regulated fusion of carrot protoplasts. In: Potrykus I, Harms CT, Hinnen A, Hütter R, King PJ, Shillito RD (eds) Protoplasts 1983 – poster proceedings. Birkhäuser, Basel, pp 96–97

Brabec F (1954) Untersuchungen über die Natur der Winklerschen Burdonen auf Grund neuen experimentellen Materials. Planta (Berl) 144:562–606
Bracha M, Sher N (1981) Fusion of enucleated protoplasts with nucleated miniprotoplasts in onion (*Allium cepa* L.). Plant Sci Lett 23:95–101
Brar DS, Ono M, Kobayashi S, Uchimiya H, Harada H (1984) Analysis of chromosomes and ribosomal RNA genes in parasexual hybrids of *Nicotiana*. Protoplasma 121:228–231
Broglie R, Coruzzi G, Fraley RT, Rogers SG, Horsch RB, Niedermeyer JG, Fink C, Flick JS, Chua NH (1984) Light regulated expression of a pea ribulose-1,5-bisphosphate carboxylase small subunit gene in transformed plant cells. Science (Wash DC) 224:838–843
Buder J (1911) Studien an *Laburnum adami*. Z Induct Abstammungs-Vererbungsl 5:209–284
Burgess J (1972) The occurrence of plasmodesma-like structures in a non-division wall. Protoplasma 74:449–458
Burgess J, Fleming EN (1974) Ultrastructural studies of the aggregation and fusion of plant protoplasts. Planta (Berl) 118:183–193
Butenko RG, Kuchko AA (1979) Physiological aspects of procurement, cultivation and hybridization of isolated potato protoplasts. Sov Plant Physiol 26:901
Butenko RG, Kuchko AA (1980) Somatic hybridization of *Solanum tuberosum* and *Solanum chacoense* by protoplast fusion. In: Ferenczy L, Farkas GL (eds) Advances in protoplast research. Pergamon, Oxford, p 293
Butterfass T (1964) Die Chloroplastenzahlen in verschiedenartigen Zellen trisomer Zuckerrüben (*Beta vulgaris* L.). Z Bot 52:46–77
Carlson PS, Smith HH, Dearing RD (1972) Parasexual interspecific plant hybridization. Proc Natl Acad Sci USA 69:2292–2294
Carr DY (1976) Plasmodesmata in growth and development. In: Cunning BES, Robards AW (eds) Intercellular communication in plants: Studies on plasmodesmata. Springer, Berlin Heidelberg New York, pp 243–288
Cella R, Carbonera D, Ladarola P (1983) Characterization of intraspecific somatic hybrids of carrot obtained by fusion of iodoacetate-inactivated A2CA resistant and sensitive protoplasts. Z Pflanzenphysiol 112:449–457
Chaleff RS, Parsons MF (1978) Isolation of a glycerol-utilizing mutant of *Nicotiana tabacum*. Genetics 89:723–728
Chen K, Wildman SG, Smith HH (1977) Chloroplast DNA distribution in parasexual hybrids as shown by polypeptide composition of fraction I protein. Proc Natl Acad Sci USA 74:5109–5112
Chetrit P, Mathieu C, Vedel F, Pelletier G, Primard C (1985) Mitochondrial DNA polymorphism induced by protoplast fusion in cruciferae. Theor Appl Genet 361:366
Chien YC, Kao KN, Wetter LR (1982) Chromosomal and isoenzyme studies of *Nicotiana tabacum-Glycine max* hybrid cell lines. Theor Appl Genet 62:301–304
Chupeau Y, Missonier C, Hommel M-C, Goujaud J (1978) Somatic hybrids of plants by fusion of protoplasts. Observations on the model system *Nicotiana glauca – Nicotiana langsdorfii*. Mol Gen Genet 165:239–245
Cocking EC (1960) A method for the isolation of plant protoplasts and vacuoles. Nature (Lond) 187:962–963
Cocking EC (1977) Uptake of foreign genetic material by plant protoplasts. In: Bourne GH, Danielli JF (eds) International review of cytology 48. Academic Press, New York, pp 323–343
Cocking EC (1978) Selection and somatic hybridisation. In: Thorpe TA (ed) Frontiers of plant tissue culture. Univ Calgary, pp 151:158
Cocking EC, George D, Price-Jones MJ, Power JB (1977) Selection procedures for the production of interspecies somatic hybrids of *Petunia hybrida* and *Petunia parodii*. II Albino complementation selection. Plant Sci Lett 10:7–12
Constabel F, Kao KN (1974) Agglutination and fusion of plant protoplasts by polyethylen glycol. Can J Bot 52:1603–1606
Constabel F, Dudits D, Gamborg OL, Kao KN (1975a) Nuclear fusion in intergeneric heterokaryons. A note. Can J Bot 53:2092–2095
Constabel F, Dudits D, Kao KN, Kartha KK (1975b) Are there no limitations to intergeneric somatic hybridization in angiosperms. Abstr 12th Int Bot Congr Leningrad, p 285

Constabel F, Weber G, Kirkpatrick JW, Pahl K (1976) Cell division of intergeneric protoplast fusion products. Z Pflanzenphysiol 79:1–7

Constabel F, Weber G, Kirkpatrick JW (1977) Sur la compatibilité des chromosomes dans les hybrides intergénériques de cellules de *Glycine max* (×) *Vicia hajastana*. CR Acad Sci Paris 285:319–322

Cséplö A, Nagy F, Maliga P (1983) Rescue of the cytoplasmic lincomycin resistance factor from *Nicotiana silvestris* into *N. plumbaginifolia* by protoplast fusion. In: Potrykus I, Harms CT, Hinnen A, Hütter R, King PJ, Shillito RD (eds) Protoplasts 1983 – poster proceedings. Birkhäuser, Basel, pp 126–127

Cséplö A, Nagy F, Maliga P (1984) Interspecific protoplast fusion to rescue a cytoplasmic lincomycin resistance mutation into fertile *Nicotiana plumbaginifolia* plants. Mol Gen Genet 198:7–11

Curtis ASG (1960) Cell contacts: some physical considerations. Am Nat 94:37–56

Davey MR, Clothier RH, Balls M, Cocking EC (1978) An ultrastructural study of the fusion of cultured amphibien cells with higher plant protoplasts. Protoplasma 96:157–172

Davey MR, Pearce N, Cocking EC (1980) Fusion of legume root nodule protoplasts with non-legume protoplasts: ultrastructural evidence for the functional activity of *Rhizobium* bacteroids in a heterokaryotic cytoplasm. Z Pflanzenphysiol 99:435–447

Dix PJ, Joo F, Maliga P (1977) Cell line of *Nicotiana sylvestris* with resistance to kanamycin and streptomycin. Mol Gen Genet 157:285–290

Dörr I (1968) Plasmatische Verbindungen zwischen artfremden Zellen. Naturwissenschaften 55:396

Douglas GC, Keller WA, Setterfield G (1981 a) Somatic hybridization between *Nicotiana rustica* and *N. tabacum*. II Protoplast fusion and selection and regeneration of hybrid plants. Can J Bot 59:220–227

Douglas GC, Wetter LR, Nakamura C, Keller WA, Setterfield G (1981 b) Somatic hybridization between *N. rustica* and *N. tabacum*. III. Biochemical, morphological and cytological analysis of somatic hybrids. Can J Bot 59:228–237

Douglas GC, Wetter LR, Keller WA, Setterfield G (1981 c) Somatic hybridization between *Nicotiana rustica* and *N. tabacum*. IV. Analysis of nuclear and chloroplast genome expression in somatic hybrids. Can J Bot 59:1509–1513

Dudits D, Kao KN, Constabel F, Gamborg OL (1976) Fusion of carrot and barley protoplasts and division of heterokaryocytes. Can J Genet Cytol 18:263

Dudits D, Hadlaczky G, Lévi E, Fejèr O, Haydki Z (1977) Somatic hybridization of *Daucus carota* and *Daucus capillifolius* by protoplast fusion. Theor Appl Genet 51:127–132

Dudits D, Hadlaczky G, Bajszar G, Koncz C, Lázár G, Horvarth G (1979) Plant regeneration from intergeneric cell hybrids. Plant Sci Lett 15:101–112

Dudits D, Fejèr O, Hadlaczky G, Koncz L, Lázár GB, Horvarth G (1980) Intergeneric gene transfer mediated by plant protoplast fusion. Mol Gen Genet 179:283–288

Evans DA, Wetter LR, Gamborg OL (1980) Somatic hybrid plants of *Nicotiana glauca* and *Nicotiana tabacum* obtained by protoplast fusion. Physiol Plant 48:225–230

Evans DA, Flick CE, Jensen RA (1981) Disease resistance: Incorporation into sexually incompatible somatic hybrids of the genus *Nicotiana*. Science (Wash DC) 213:907–909

Evans DA, Flick CE, Kut SA, Reed SM (1982) Comparison of *Nicotiana tabacum* and *Nicotiana nesophila* hybrids produced by ovule culture and protoplast fusion. Theor Appl Genet 62:193–197

Evans DA, Bravo JE, Kut SA, Flick CE (1983) Genetic behaviour in somatic hybrids in the genus *Nicotiana: N. otophora* (×) *N. tabacum* and *N. sylvestris* (×) *N. tabacum*. Theor Appl Genet 64:93–101

Evola SV (1983) Chlorate resistant variants of *Nicotiana tabacum* L. II. Parasexual genetic characterization. Mol Gen Genet 189:441–446

Evola SV, Earle ED, Chaleff RS (1983) The use of genetic markers selected in vitro for the isolation and verification of intraspecific somatic hybrids of *Nicotiana tabacum* L. Mol Gen Genet 189:441–446

Fankhauser H, Gebhardt C, Jia JF, King PJ, Laser M, Lázár G, Potrykus I, Shillito R, Shimamoto K (1983) Fusion complementation tests on a group of independently-isolated auxotrophic and temperature-sensitive clones of *Hyoscyamus muticus* and *Nicotiana tabacum*. In:

Potrykus I, Harms CT, Hinnen A, Hütter R, King PJ, Shillito RD (eds) Protoplasts 1983 – poster proceedings. Birkhäuser, Basel, pp 110–111
Ferenczy L, Kevei F, Zsolt J (1974) Fusion of fungal protoplasts. Nature (Lond) 248:793–794
Flick CE, Evans DA (1982) Evaluation of cytoplasmic segregation in somatic hybrids of *Nicotiana:* tentoxin sensitivity. J Hered 73:264–266
Fluhr R, Aviv D, Edelman M, Galun E (1983) Cybrids containing mixed and sorted-out chloroplasts following intraspecific somatic fusions in *Nicotiana*. Theor Appl Genet 65:289–294
Fluhr R, Aviv D, Galun E, Edelman M (1984) Generation of heteroplastidic *Nicotiana* cybrids by protoplast fusion: analysis for plastid recombinant types. Theor Appl Genet 67:491–497
Fowke LC, Gamborg OL (1980) Applications of protoplasts to the study of plant cells. Int Rev Cytol 68:9–48
Fowke LC, Bech-Hansen CW, Gamborg OC, Constabel F (1975a) Electron-microscope observations of mitosis and cytokinesis in multinucleate protoplasts of soybean. J Cell Sci 18:491–507
Fowke LC, Rennie PJ, Kirkpatrick JW, Constabel F (1975b) Ultrastructural characteristics of intergeneric protoplast fusion. Can J Bot 53:272–278
Fowke LC, Rennie PJ, Kirkpatrick JW, Constabel F (1976) Ultrastructure of fusion products from soybean cell culture and sweet clover leaf protoplasts. Planta (Berl) 130:39–45
Fowke LC, Constabel F, Gamborg OL (1977) Fine structure of fusion products from soybean cell culture and pea leaf protoplasts. Planta (Berl) 135:257–266
Frankel R (1971) Genetical evidence on alternative maternal and mendelian heredity elements in *Petunia hybrida*. Heredity 26:107–119
Frankel R, Galun E (1977) Pollination mechanisms, reproduction and plant breeding. Monographs on Theoretical and Applied Genetics. Vol 2. Springer, Berlin Heidelberg New York
Galbraith DW (1984) Selection of somatic hybrid cells by fluorescence-activated cell sorting. In: Vasil IK (ed) Cell culture and somatic cell genetics. V 1 – Laboratory procedures and their applications. Academic Press, Orlando, pp 433–447
Galbraith DW, Galbraith JEC (1979) A method for identification of fusion of plant protoplasts derived from tissue culture. Z Pflanzenphysiol 93:149–158
Galbraith DW, Harkins KR (1982) Cell sorting as a means for isolating somatic hybrids. In: Fujiwara A (ed) Plant tissue culture 1982. Maruzen, Tokyo, pp 617–620
Galbraith DW, Mauch TJ (1980) Identification of fusion of plant protoplast. II Conditions for the reproducible fluorescence labelling of protoplasts derived from mesophyll tissue. Z Pflanzenphysiol 48:129–140
Galbraith DW, Afonso CL, Harkins KR (1984) Flow sorting and culture of protoplasts: conditions for high-frequency recovery, growth and morphogenesis from sorted protoplasts of suspension cultures of Nicotiana. Plant Cell Rep 3:151–155
Galun E, Arzee-Gonen P, Fluhr R, Edelman M, Aviv D (1982a) Cytoplasmic hybridization in *Nicotiana:* Mitochondrial DNA analysis in progenies resulting from fusion between protoplasts having different organelle constitutions. Mol Gen Genet 186:50–56
Galun E, Bleichman S, Aviv D (1982b) Development of an organelle-genetics system by unilateral transfer of organelles through protoplast fusion and cybrid formation. In: Fujiwara A (ed) Plant tissue culture 1982. Maruzen, Tokyo, pp 645–648
Galun E, Arzee-Gonen P, Bleichman S, Fluhr R, Edelman M, Aviv D (1983) Identification of plastomes and chondriomes of somatic hybrid plants resulting from protoplast by molecular probing of fractionated DNA. In: Potrykus I, Harms CT, Hinnen A, Hütter R, King PJ, Shillito RD (eds) Protoplasts 1983 – poster proceedings. Birkhäuser, Basel, pp 118–119
Gamborg OL, Shyluk JP, Shahin EA (1981) Isolation, fusion and culture of protoplasts. In: Thorpe TA (ed) Plant tissue culture – methods and applications in agriculture. Academic Press, New York, pp 115–153
Gengenbach BG, Connelly JA, Pring DR, Conde MF (1981) Mitochondrial DNA variation in maize plants regenerated during tissue culture selection. Theor Appl Genet 59:161–168
Gleba YY (1979) Nonchromosomal inheritance in higher plants as studied by somatic cell hybridization. In: Sharp WR, Larsen PO, Raghavan V (eds) Plant cell and tissue culture – principles and applications. Ohio State Univ Press, Columbus, pp 775–788

Gleba YY (1980) Protoplast fusion and genetic engineering of higher plants. Habil Thesis, Leningrad

Gleba YY, Berlin J (1979) Somatic hybridization by protoplast fusion in *Nicotiana:* fate of nuclear genetic determinants. Abstr fifth int protopl symp. Szeged, Hungary: 73

Gleba YY, Hoffmann F (1978) Hybrid cell lines *Arabidopsis thaliana* (×) *Brassica campestris:* No evidence for specific chromosome elimination.Mol Gen Genet 165:257–264

Gleba YY, Hoffmann F (1979) *Arabidobrassica:* Plant genome engineering by protoplast fusion. Naturwissenschaften 66:547–554

Gleba YY, Hoffmann F (1980) *Arabidobrassica:* A novel plant obtained by protoplast fusion. Planta (Berl) 149:112–117

Gleba YY, Sytnik KM (1984) Protoplast fusion – genetic engineering in higher plants. Monographs on theoretical and applied genetics, vol 8. Springer, Berlin Heidelberg New York Tokyo

Gleba YY, Butenko RG, Sytnik KM (1975) Fusion of protoplasts and parasexual hybridization in *Nicotiana tabacum.* Dokl Akad Nauk SSSR 221:1196–1198

Gleba YY, Kohlenbach HW, Hoffmann F (1978) Root morphogenesis in somatic hybrid cell lines *Arabidopsis thaliana* (×) *Brassica campestris.* Naturwissenschaften 65:655–656

Gleba YY, Momot VP, Cherep NN, Skarzynskaya MV (1982) Intertribal hybrid cell lines of *Atropa belladonna* (×) *Nicotiana chinensis* obtained by cloning individual protoplast fusion products. Theor Appl Genet 62:75–79

Gleba YY, Momot VP, Okolot AN, Cherep NN, Skarzynskaya MV, Kotov V (1983) Genetic process in intergeneric cell hybrids *Atropa* (×) *Nicotiana.* 1. Genetic constitutions of cells of different clonal origin grown in vitro. Theor Appl Genet 65:269–276

Gleba YY, Komarnitzky IK, Kolesnik NN, Meshkene I, Mortyn GI (1985) Transmission genetics of the somatic hybridization process in *Nicotiana.* II. Plastome heterozygotes. Mol Gen Genet 198:476–481

Gleddie S, Keller WA, Setterfields G, Wetter LR (1983) Somatic hybridization between *Nicotiana rustica* and *N. sylvestris.* Plant Cell Tissue Organ Cult 2:269–283

Glimelius K, Bonnett HT (1981) Somatic hybridization in *Nicotiana:* Restoration of photoautotrophy to an albino mutant with defective plastids. Planta (Berl) 153:497–503

Glimelius K, Eriksson T, Grafe R, Müller AJ (1978) Somatic hybridization of nitrate reductase-deficient mutants of *Nicotiana tabacum* by protoplast fusion. Physiol Plant 44:273–277

Glimelius K, Chen K, Bonnett H (1981) Somatic hybridization in *Nicotiana:* segregation of organellar traits among hybrid and cybrid plants. Planta (Berl) 153:504–510

Gosch G, Reinert J (1976) Nuclear fusion in intergeneric heterokaryocytes and subsequent mitosis of hybrid nuclei. Naturwissenschaften 63:534–535

Gosch G, Reinert J (1978) Cytological identification of intergeneric somatic hybrid cells. Protoplasma 96:23–38

Grafe R, Müller AJ (1983) Complementation analysis of nitrate reductase deficient mutants of *Nicotiana tabacum* by somatic hybridization. Theor Appl Genet 66:127–130

Gray JC, Wildman SG (1976) A specific immunoabsorbant for the isolation of fraction I protein. Plant Sci Lett 6:91–96

Gressel J, Cohen N, Binding H (1984) Somatic hybridization of an atrazine resistant biotype of *Solanum nigrum* with *Solanum tuberosum.* 2. Segregation of plastomes. Theor Appl Genet 67:131–134

Grimsley NH, Ashton NW, Cove DH (1977a) The production of somatic hybrids by protoplast fusion in the moss *Physcomitrella patens.* Mol Gen Genet 154:97–100

Grimsley NH, Ashton NW, Cove DH (1977b) Complementation analysis of auxotrophic mutants of the moss *Physcomitrella patens,* using protoplast fusion. Mol Gen Genet 155:103–107

Gupta SB, Gupta P (1973) Selective elimination of *Nicotiana glutinosa* chromosomes in the F1 hybrids of *N. suaveolens* and *N. glutinosa.* Genetics 73:605–612

Gupta PP, Gupta M, Schieder O (1982) Correction of nitrate reductase defect in auxotrophic plant cells through protoplast-mediated intergeneric gene transfers. Mol Gen Genet, pp 378–383

Gupta PP, Schieder O, Gupta M (1984) Intergeneric nuclear gene transfer between somatically and sexually incompatible plants through asymmetric protoplast fusion. Mol Gen Genet 197:30–35

Hämmerling J (1963) Nucleus- cytoplasmic interactions in *Acetabularia* and other cells. Annu Rev Plant Physiol 14:65–92
Harkins KR, Galbraith DW (1984) Flow sorting and culture of plant protoplasts. Physiol Plant 60:43–52
Harms CT (1977) In vitro-Kultur von Mais und Tabak und Entwicklung eines Gradientensystems für Frühselektion fusionierter Protoplasten. Dissertation, Heidelberg
Harms CT (1983) Somatic incompatibility in the development of higher plant somatic hybrids. Quart Rev Biol 58:325–353
Harms CT, Oertli J (1982) Complementation and expression of amino acid analogue resistance studied by intraspecific and interfamily protoplast fusion. In: Fujiwara A (ed) Plant Tissue Culture 1982. Maruzen, Tokyo, pp 467–468
Harms CT, Potrykus I (1978) Enrichment for heterokaryocytes by the use of isoosmotic density gradients after plant protoplast fusion. Theor Appl Genet 53:49–55
Harms CT, Potrykus I (1979) Induced buoyant density shift of protoplasts and somatic hybridization studies with *Citrus* and tobacco. In: Ferenczy L, Farkas GL (eds) Advances in protoplast research. Pergamon Press, Oxford
Harms CT, Potrykus I, Widholm JM (1981) Complementation and dominant expression of amino acid analogue resistance markers in somatic hybrid clones from *Daucus carota* after protoplast fusion. Z Pflanzenphysiol 101:377–390
Harms CT, Oertli J, Widholm JM (1982) Characterization of amino acid analogue resistant somatic hybrid cell lines of *Daucus carota* L. Z Pflanzenphysiol 106:239–249
Hauptmann R, Kumar P, Widholm J (1983) Carrot (×) tobacco somatic cell hybrids selected by amino acid analog resistance complementation. In: Potrykus I, Harms CT, Hinnen A, Hütter R, King PJ, Shillito RD (eds) Protoplasts 1983 – poster proc. Birkhäuser, Basel, pp 92–93
Hein T, Przewozny T, Schieder O (1982) The use of an auxotrophic cell line for the culture and selection of somatic hybrids. Theor Appl Genet 64:119–122
Hein T, Przewozny T, Schieder O (1983) Culture and selection of somatic hybrids using an auxotrophic cell line. Theor Appl Genet 64:119–122
Herrera-Estrella L, Van den Broeck G, Maenhaut R, Van Montagu M, Schell J, Timko M, Cashmore A (1984) Light inducible and chloroplast-associated expression of a chimeric gene introduced into *Nicotiana tabacum* using a Ti plasmid vector. Nature (Lond) 310:115–120
Ho KM, Kasha KJ (1975) Genetic control of chromosome elimination during haploid formation in barley. Genetics 81:263–275
Hodgson RAJ, Rose RJ (1984) Fusion of spinach mesophyll protoplasts with carrot root parenchyma protoplasts and the effect of spinach chloroplasts. J Plant Physiol 115:69–78
Hoffmann F, Adachi T (1981) *Arabidobrassica:* Chromosomal recombination and morphogenesis in asymmetric intergeneric hybrid cells. Planta (Berl) 153:586–593
Hofmeister L (1954) Mikrurgische Untersuchung über die geringe Fusionsneigung plasmolysierter, nackter Protoplasten. Protoplasma 43:278–326
Horn ME, Kameya T, Brotherton JE, Widholm JM (1983) The use of amino acid analog resistance and plant regeneration ability to select somatic hybrids between *Nicotiana tabacum* and *N. glutinosa.* Mol Gen Genet 192:235–240
Ito M, Maeda M (1973) Fusion of meiotic protoplasts in Liliaceous plants. Exp Cell Res 80:453–456
Iwai S, Nagao T, Nakata K, Kawashima N, Matsuyama S (1980) Expression of nuclear and chloroplastic genes coding for fraction-1 protein in somatic hybrids of *Nicotiana tabacum* (×) *rustica.* Planta (Berl) 147:414–417
Iwai S, Nakata K, Nagao I, Kawashima N, Mutsuyama S (1981) Detection of the *Nicotiana rustica* chloroplast genome coding for the large subunit of fraction I protein in a somatic hybrid in which only the *N. tabacum* chloroplast genome appeared to have been expressed. Planta (Berl) 152:478–480
Izhar S, Power JB (1979) Somatic hybridization in *Petunia:* a male sterile cytoplasmic hybrid. Plant Sci Lett 14:49–55
Izhar S, Tabib Y (1980) Somatic hybridization in *Petunia.* II. Heteroplasmic state in somatic hybrids followed by cytoplasmic segregation into male sterile and male fertile lines. Theor Appl Genet 57:241–245

Izhar S, Schlicter M, Swartzberg D (1983) Sorting out of cytoplasmic elements in somatic hybrids of *Petunia* and the prevalence of the heteroplasmon through several meiotic cycles. Mol Gen Genet 190:459–467

Izhar S, Tabib Y, Swartzberg D (1984) Reciprocal transfer of male sterile and normal plasmons in *Petunia*. Theor Appl Genet 68:455–457

Jia J, Potrykus I, Lázár GB, Saul M (1983) *Hyoscyamus-Nicotiana* fusion hybrids selected via auxotroph complementation and verified by species-specific DNA hybridization. In: Potrykus I, Harms CT, Hinnen A, Hütter R, King PJ, Shillito RD (eds) Protoplasts 1983 – poster proc. Birkhäuser, Basel, pp 110–111

Jones MGK (1976) The origin and development of plasmodesmata. In: Gunning BES, Robards AW (eds) Intercellular communication in plants. Studies on plasmodesmata. Springer, Berlin Heidelberg New York, pp 81–105

Jones CW, Mastrangelo IA, Smith HH, Liu HZ, Meck RA (1976) Interkingdom fusion between human (HeLa) cells and tobacco hybrid (GGLL) protoplasts. Science (Wash DC) 193:401–403

Kameya T (1979) Studies on plant cell fusion: effects of dextran and pronase E on fusion. Cytologia (Tokyo) 44:449–456

Kameya T (1982) The method for fusion with dextran. In: Fujiwara A (ed) Plant tissue culture 1982. Maruzen, Tokyo, pp 613–616

Kameya T, Horn ME, Widholm JM (1981) Hybrid shoot formation from fused *Daucus carota* and *Daucus capillifolius* protoplasts. Z Pflanzenphysiol 104:459–466

Kao KN (1977) Chromosomal behaviour in somatic hybrids of soybean-*Nicotiana glauca*. Mol Gen Genet 150:225–230

Kao KN, Michayluk MR (1974) A method for high-frequency intergeneric fusion of plant protoplasts. Planta (Berl) 115:355–367

Kao KN, Constabel F, Michayluk MR, Gamborg OL (1974) Plant protoplast fusion and growth of intergeneric hybrid cells. Planta (Berl) 120:215–227

Kartha KK, Gamborg OL, Constabel F, Kao KN (1974) Fusion of rapeseed and soybean protoplasts and subsequent division of heterokaryocytes. Can J Bot 52:2435–2436

Kasha KJ, Kao KN (1970) High frequency haploid production in barley (*Hordeum vulgare* L.). Nature (Lond) 225:874–876

Keller WA, Melchers G (1973) The effect of high pH and calcium on tobacco leaf protoplast fusion. Z Naturforsch Sect C Biosci 28:737–741

Klercker af J (1892) Eine Methode zur Isolierung lebender Protoplasten. Öfvers Kongl Akad Förh, Stockholm 9:463–471

Kinsara AM, Cocking EC (1983) Assessment of somatic hybridization between *Lycopersicon esculentum* and *L. peruvianum*. In: Potrykus I, Harms CT, Hinnen A, Hütter R, King PJ, Shillito RD (eds) Protoplasts 1983 – poster proc. Birkhäuser, Basel, pp 80–81

Kollmann R, Dörr I (1969) Strukturelle Grundlagen des zwischenzelligen Stoffaustausches. Ber Dtsch Bot Ges 82:415–425

Kollmann R, Glockmann C (1985) Studies on graft unions. I. Plasmodesmata between cells of plants belonging to different unrelated taxa. Protoplasma 124:224–235

Kollmann R, Yang S, Glockmann C (1985) Studies on graft unions. II. Continuous and half plasmodesmata in different regions of the graft interface. Protoplasma 126:19–29

Komarnitzky IK, Kuchko AA, Butenko RG, Vitenko VA (1980) Ribulose diphosphate carboxylase of somatic hybrid of potato. Dokl Akad Nauk USSR 4:75–78

Komarnitzky IK, Kuchko AA, Shlumukov LR, Butenko RG (1981) Analysis of chl-DNA of interspecific somatic hybrid of potato. Dokl Akad Nauk USSR 256:495–496

Koop HU (1984) Entwicklung elektrisch fusionierter Protoplasten in individueller Kultur. In: Mitteilungsband – Kurzfassungen der Beiträge, Botaniker-Tagung in Wien 1984. Inst für Bot, Univ Wien, S 72

Koop HU, Dirk J, Wolff D, Schweiger HG (1983) Somatic hybridization of two selected single cells. Cell Biol Int Rep 7:1123–1128

Krumbiegel G (1980) Versuche zur sexuellen und somatischen Hybridisierung von *Atropa belladonna* L. und *Datura innoxia* Mill. Diss, Köln

Krumbiegel G, Schieder O (1979) Selection of somatic hybrids after fusion of protoplasts from *Datura innoxia* Mill and *Atropa belladonna* L. Planta (Berl) 145:371–375

Krumbiegel G, Schieder O (1981) Comparison of somatic and sexual incompatibility between *Datura innoxia* and *Atropa belladonna*. Planta (Berl) 153:466–470

Kumar A, Cocking EC, Bovenberg WA, Kool AJ (1982) Restriction endonuclease analysis of chloroplast DNA in interspecific somatic hybrids of Petunia. Theor Appl Genet 62:377–383

Küster E (1910) Eine Methode zur Gewinnung abnorm großer Protoplasten. Arch Entwicklungsmech Org (Wilhelm Roux) 30:351–355

Küster E (1958) Plasmoptyse. In: Heilbrunn LV, Weber F (eds) Protoplasmatologia, Handbuch der Protoplasmaforschung II, Cytoplasma C7b. Springer, Wien, S 1–39

Lázár GB, Dudits D, Sung ZR (1981) Expression of cycloheximide resistance on carrot somatic hybrids and their segregants. Genetics 98:347–356

Lázár GB, Fankhäuser H, Potrykus I (1983) Complementation analysis of a nitrate reductase deficient *Hyoscyamus muticus* cell line by somatic hybridization. Mol Gen Genet 189:359–364

Le Baron H, Gressel J (1982) Herbicide resistance in plants. Wiley, New York

Leible MB, Shoeman RL, Schweiger HG (1982) Ribulose-1,5-bisphosphate carboxylase, a marker for chloroplast species specifity in *Acetabularia*. Biochim Biophys Acta 699:60–66

Lemieux C, Turmel M, Lee RW (1981) Physical evidence for recombination of chloroplast DNA in hybrid progeny of *Chlamydomonas eugametos* and *C. moewusii*. Current Genet 3:97–103

Li X, Schieder O, Meijuan H, Li W (1983) The transfer of LpDH activity as marker in somatic hybrid plants between tobacco tumor B6S3 and normal tobacco xanthi. In: Potrykus I, Harms CT, Hinnen A, Hütter R, King PJ, Shillito RD (eds) Protoplasts 1983 – poster proc. Birkhäuser, Basel, pp 86–87

Littlefield JW (1964) Selection of hybrids from mating of fibroblasts *in vitro* and their presumed recombinants. Science (Wash DC) 145:709–710

Lönnendonker N, Schieder O (1980) Amylase isoenzymes of the genus *Datura* as a simple method for an early identification of somatic hybrids. Plant Sci Lett 17:135–139

Lörz H (1984) Isolated cell organelles and subprotoplasts – their role in somatic cell genetics. In: Dodds JH (ed) Plant genetic engineering. Cambridge Univ Press, pp 27–59

Lörz H, Potrykus I (1980) Isolation of subprotoplasts for genetic manipulation studies. In: Ferenczy L, Farkas GL (eds) Advances in protoplast research. Pergamon, Oxford, pp 377–382

Lörz H, Paszkowski J, Dierks-Ventling C, Potrykus I (1981) Isolation and characterization of cytoplasts and miniprotoplasts derived from protoplasts of cultured cells. Physiol Plant 53:385–391

Maliga P, Lázár G, Joó F, Nagy AH, Menczel L (1977) Restoration of morphogenetic potential in *Nicotiana* by somatic hybridization. Mol Gen Genet 157:291–296

Maliga P, Kiss ZR, Nagy AH, Lázár G (1978) Genetic instability in somatic hybrids of *Nicotiana tabacum* and *Nicotiana knightiana*. Mol Gen Genet 163:145–152

Maliga P, Lörz H, Lázár G, Nagy F (1982) Cytoplast-protoplast fusion for interspecific chloroplast transfer in *Nicotiana*. Mol Gen Genet 185:211–215

Márton L, Sidorov V, Biasini G, Maliga P (1982) Complementation in somatic hybrids indicated four types of nitrate reductase deficient lines in *Nicotiana plumbaginifolia*. Mol Gen Genet 187:1–3

Márton L, Biasini G, Sidorov V, Maliga P (1983) Nitrate-reductase deficients in the progeny after selfing complemented somatic hybrids of *Nicotiana plumbaginifolia*. In: Potrykus I, Harms CT, Hinnen A, Hütter R, King PJ, Shillito RD (eds) Protoplasts 1983 – poster proc. Birkhäuser, Basel, pp 100–101

Márton L, Biasini G, Maliga P (1985) Co-segregation of nitrate-reductase activity and normal regeneration ability in selfed sibs of *Nicotiana plumbaginifolia* somatic hybrids, heterozygotes for nitrate-reductase deficiency. Theor Appl Genet 70:340–344

Medgyesy P, Menczel L, Maliga P (1980) The use of cytoplasmic streptomycin resistance: chloroplast transfer from *Nicotiana tabacum* into *Nicotiana sylvestris*, and isolation of their somatic hybrids. Mol Gen Genet 179:693–698

Medgyesy P, Nagy F, Menczel L, Maliga P (1983) Selection for cytoplasmic streptomycin resistance after protoplast fusion as a tool for transfer of cytoplasmic male sterility / cms / in *Nicotiana*. In: Portykus I, Harms CT, Hinnen A, Hütter R, King PJ, Shillito RD (eds) Protoplasts 1983 – poster proc. Birkhäuser, Basel, pp 124–125

Medgyesy P, Fejer E, Maliga P (1985) Interspecific chloroplast recombination in a *Nicotiana* somatic hybrid. Abstr Int Symp Genetic Manipulation in Plant Breeding. Berlin, p 86

Melchers G (1977) Microbial techniques in somatic hybridization by fusion of protoplasts. In: Brinkley BR, Porter KR (eds) International cell biology. The Rockefeller Univ Press, pp 207–215

Melchers G, Labib G (1974) Somatic hybridization of plants by fusion of protoplasts. I. Selection of light resistant hybrids of "haploid" light sensitive varieties of tobacco. Mol Gen Genet 135:277–294

Melchers G, Sacristán MD, Holder AA (1978) Somatic hybrid plants of potato and tomato regenerated from fused protoplasts. Carlsberg Res Commun 43:203–218

Menczel L, Lázár G, Maliga P (1978) Isolation of somatic hybrids by cloning *Nicotiana* heterokaryons in nurse culture. Planta (Berl) 143:29–32

Menczel L, Nagy F, Kiss ZS, Maliga P (1981) Streptomycin resistant and sensitive somatic hybrids of *Nicotiana tabacum* (×) *Nicotiana knightiana:* correlation of resistance to *N. tabacum* plastids. Theor Appl Genet 59:191–195

Menczel L, Nagy F, Lázár G, Maliga P (1983) Transfer of cytoplasmic male sterility by selection for streptomycin resistance after protoplast fusion in *Nicotiana.* Mol Gen Genet 189:365–369

Michaelis P (1955) Über Gesetzmäßigkeiten der Plasmon-Umkombination und über eine Methode zur Trennung einer Plastiden-, Chondriosomen – resp. Sphaerosomen – (Mikrosomen-) und einer Zytoplasmavererbung. Cytologia 20:315–388

Miller RA, Gamborg OL, Keller WA, Kao KN (1971) Fusion and division of nuclei in multinucleated soybean protoplasts. Can J Genet Cytol 13:347–353

Moore R (1983) Physiological aspects of graft formation. In: Moore R (ed) Vegetative compatibility responses in plants. Baglor Univ Press, Waco, pp 89–121

Müller AG, Grafe R (1978) Isolation and characterization of cell lines of *Nicotiana tabacum* lacking nitrate reductase. Mol Gen Genet 161:67–76

Müller-Gensert E, Landsmann J, Eckes P, Schieder O (1983) Confirmation of chloroplast segregation in somatic hybrids of *Datura* by DNA-DNA-hybridization. In: Potrykus I, Harms CT, Hinnen A, Hütter R, King PJ, Shillito RD (eds) Protoplasts 1983 – poster proc. Birkhäuser, Basel, pp 106–107

Nagao T (1978) Somatic hybridization by fusion of protoplasts. I. The combination of *Nicotiana tabacum* and *Nicotiana rustica.* Jpn J Crop Sci 48:385

Nagao T (1979) Somatic hybridization by fusion of protoplasts. II. The combination of *Nicotiana tabacum* and *N. glutinosa* and *N. tabacum* and *N. alata.* Jpn J Crop Sci 47:491

Nagao T (1982) Somatic hybridization by fusion of protoplasts: III. Somatic hybrids of sexually incompatible combinations *Nicotiana tabacum* (×) *Nicotiana repanda* and *Nicotiana tabacum* (×) *Salpiglossis sinuata.* Jpn J Crop Sci 51:35–42

Nagata T, Melchers G (1978) Surface charge of protoplasts and their significance in cell-cell interaction Planta (Berl) 142:235–238

Nagata T, Eibel H, Melchers G (1979) Fusion of plant protoplasts induced by a positively charged synthetic phospholipid. Z Naturforsch Sect C Biosci 34:460–462

Nagl W, Hoffmann F (1980) *Arabidobrassica:* Evidence for intergeneric somatic hybrid nature from electron microscopic morphometry of chromatin. Eur J Cell Biol 21:227–228

Nakata K, Oshima H (1982) Cytoplasmic chimaericity in the somatic hybrids of tobacco. In: Fujiwara A (ed) Plant tissue culture 1982. Maruzen, Tokyo, pp 641–642

Nagy F, Török I, Maliga P (1981) Extensive rearrangements in the mitochondrial DNA in somatic hybrids of *Nicotiana tabacum* and Nicotiana knightiana. Mol Gen Genet 183:437–439

Nagy F, Lázár G, Menczel L, Maliga P (1983) A heteroplasmic state induced by protoplast fusion is a necessary condition for detecting rearrangements in *Nicotiana* mitochondrial DNA. Theor Appl Genet 66:203–207

Nehls R (1978) The use of metabolic inhibitors for the selection of fusion products of higher plant protoplasts. Mol Gen Genet 166:117–118

Nehls R (1981) Versuche zur Regeneration und Hybridisierung von Protoplasten höherer Pflanzen und zur Selektion von Fusionsprodukten. Diss Kiel

Nehls R, Binding H (1979) Biochemical selection of fusion products from higher plant protoplasts. Abstr 5th int protoplast symp. Szeged, p 98

Ninnemann H, Jüttner F (1981) Volatile substances from tissue cultures of potato, tomato and their somatic fusion products. – Comparison of gas chromatographic patterns for identification of hybrids. Z Pflanzenphysiol 103:95–108

O'Connell PJ, Brady CJ (1981) Multiple forms of the large subunit of wheat ribulose bisphosphate generated by excess iodoacetamide. Biochim Biophys Acta 670:355–361

Ootaki T, Kinno T, Yoshida K (1977) Complementation between *Phycomyces* mutants of mating type (+) with abnormal phototropism. Mol Gen Genet 152:245–251

Patnaik G, Cocking EC, Hamill J, Pental D (1982) A simple procedure for the manual isolation and identification of plant heterokaryons. Plant Sci Lett 24:105–110

Pelletier G, Chupeau Y (1984) Plant protoplast fusion and somatic plant cell genetics (review). Physiol Vég 22:377–399

Pelletier G, Primard C, Vedel F, Chetrit P, Remy P, Rousselle P, Renard M (1983) Intergeneric cytoplasmic hybridization in cruciferae by protoplast fusion. Mol Gen Genet 191:244–250

Poste G, Allison AC (1971) Membrane fusion reaction: a theory. J Theor Biol 32:165–184

Poste G, Allison AC (1973) Membrane fusion. Biochim Biophys Acta 300:421–465

Potrykus I (1971) Intra- and interspecific fusion of protoplasts from petals of *Torenia*. Nature New Biol (Lond) 231:57–58

Potrykus I, Jia J, Lázár GB, Saul M (1984) *Hyoscyamus muticus* (×) *Nicotiana tabacum* fusion hybrids selected via auxotroph complementation. Plant Cell Rep 3:68–71

Poulsen C, Porath D, Sacristán MD, Melchers G (1980) Peptide mapping of the ribulose bisphosphate carboxylase small subunit from the somatic hybrid of tomato and potato. Carlsberg Res Commun 45:249–267

Power JB, Chapman JV (1983) Intra- and intersubfamilial somatic hybridization within the Solanaceae. In: Potrykus I, Harms CT, Hinnen A, Hütter R, King PJ, Shillito RD (eds) Protoplasts 1983 – poster proc. Birkhäuser, Basel, pp 276–277

Power JB, Frearson EM (1973) The inter- and intraspecific fusion of plant protoplasts; subsequent development in culture with reference to crown gall callus and tobacco and *Petunia* leaf system. In: Protoplasts et fusion de cellules somatiques végétales. CNRS 212, Paris, pp 409–414

Power JB, Cummins SE, Cocking EC (1970) Fusion of isolated plant protoplasts. Nature (Lond) 225:1016–1018

Power JB, Frearson EM, Hayward C, Cocking EC (1975) Some consequences of the fusion and selective culture of *Petunia* and *Parthenocissus* protoplasts. Plant Sci Lett 5:197–207

Power JB, Frearson EM, Hayward C, George D, Evans PK, Berry SF, Cocking EC (1976) Somatic hybridization of *Petunia hybrida* and *Petunia parodii*. Nature (Lond) 263:500–502

Power JB, Berry SF, Frearson EM, Cocking EC (1977) Selection procedures for the production of inter-species somatic hybrids of *Petunia hybrida* and *Petunia parodii*. I. Nutrient media and drug sensitivity complementation selection. Plant Sci Lett 10:1–6

Power JB, Berry SF, Chapman JV, Cocking EC, Sink KC (1979) Somatic hybrids between unilateral cross-incompatible *Petunia* species. Theor Appl Genet 55:97–99

Power JB, Berry SF, Chapman JV, Cocking EC (1980) Somatic hybridization of sexually incompatible Petunias: *Petunia parodii, Petunia parviflora*. Theor Appl Genet 57:1–4

Raghavan V (1977) Applied aspects of embryo culture. In: Reinert J, Bajaj YPS (eds) Applied and fundamental aspects of plant cell, tissue, and organ culture. Springer, Berlin Heidelberg New York, pp 375–397

Reinert J, Gosch G (1976) Continuous division of heterokaryons from *Daucus carota* and *Petunia hybrida* protoplasts. Naturwissenschaften 63:534

Rennie PJ, Weber G, Constabel F, Fowke LC (1980) Dedifferentiation of chloroplasts in interspecific and homospecific protoplast fusion products. Protoplasma 103:253–262

Robertson D, Earle E (1985) Synthesis of atrazine resistant *Brassica napus* through protoplast fusion. Abstr Int Symp Biotechnology in Plant Science: relevance to agriculture in the eighties. Ithaca, New York. Cornell University Biotechnology Program, p 30

Ruesink AW (1971) The plasma membrane of *Avena* coleoptile protoplasts. Plant Physiol (Bethesda) 47:192–195

Saul MW, Potrykus I (1983) Species specific DNA used to identify interspecific somatic hybrids. In: Potrykus I, Harms CT, Hinnen A, Hütter R, King PJ, Shillito RD (eds) Protoplasts 1983 – poster proc. Birkhäuser, Basel, pp 108–109

Saul MW, Potrykus I (1984) Species-specific repetitive DNA used to identify interspecific somatic hybrids. Plant Cell Rep 3:65–67

Scandalios JG (1974) Isozymes in development and differentiation. Annu Rev Plant Physiol 25:225–258

Schenck HR (1982) *Brassica napus* – successful resynthesis by protoplast fusion between *B. oleracea* and *B. campestris*. In: Fujiwara A (ed) Plant tissue culture 1982. Maruzen, Tokyo, pp 639–640

Schenck HR (1983) Selection of somatic hybrids by fusion of protoplasts from *Brassica oleracea* and *B. campestris*. In: Potrykus I, Harms CT, Hinnen A, Hütter R, King PJ, Shillito RD (eds) Protoplasts 1983 – poster proc. Birkhäuser, Basel, pp 280–281

Schenck HR, Röbbelen G (1982) Somatic hybrids by fusion of protoplasts from *Brassica oleracea* and *B. campestris*. Z Pflanzenzücht 89:278–288

Schieder O (1974) Fusionen zwischen Protoplasten von *Sphaerocarpos donnellii* Aust-Mutanten. Biochem Physiol Pflanzen 165:433–435

Schieder O (1976) Isolation of mutants with altered pigments after irradiating haploid protoplasts from *Datura innoxia* Mill with X-rays. Mol Gen Genet 149:251–254

Schieder O (1977) Hybridization experiments with protoplasts from chlorophyll-deficient mutants of some Solanaceous species. Planta (Berl) 137:253–257

Schieder O (1978) Somatic hybrids of *Datura innoxia* Mill (×) *Datura discolor Bernh.* and of *Datura innoxia* Mill (×) *Datura stramonium* L. var *tatula*. 1. Selection and characterization. Mol Gen Genet 162:113–119

Schieder O (1980c) Somatic hybrids between a herbaceous and two tree *Datura* species. Z Pflanzenphysiol 98:119–127

Schieder O (1980b) Somatic hybrids of *Datura innoxia* Mill. (×) *Datura discolor* Bernh. and of *Datura innoxia* Mill (×) *Datura stramonium* L. var. *tatula* L. II. Analysis of progenies of three sexual generations. Mol Gen Genet 179:387–390

Schieder O (1982) Somatic hybridization: a new method for plant improvement. In: Vasil I, Scowcroft WR, Frey KJ (eds) Plant improvement and somatic cell genetics. Academic Press, New York, pp 239–253

Schieder O, Vasil IK (1980) Protoplast fusion and somatic hybridization. Int Rev Cytol Suppl 11B:21–46

Schiller B, Herrmann RG, Melchers G (1982) Restriction endonuclease analysis of plastid DNA from tomato, potato and some of their somatic hybrids. Mol Gen Genet 186:453–459

Scowcroft WR, Larkin PJ (1981) Chloroplast DNA assorts randomly in intraspecific somatic hybrids of *Nicotiana debneyi*. Theor Appl Genet 60:179–184

Senda M, Morikawa H, Katagi H, Takada T, Yamada Y (1980) Effect of temperature on membrane fluidity and protoplast fusion. Theor Appl Genet 57:33–36

Senda M, Morikawa H, Takeda J (1982) Electrical induction of cell fusion of plant protoplasts. In: Fujiwara A (ed) Plant tissue culture 1982. Maruzen, Tokyo, pp 615–661

Shepard JF, Bidney D, Shahin E (1980) Potato protoplasts in crop improvement. Science (Wash DC) 208:17–24

Shepard JF, Bidney D, Barsby T, Kemble R (1983) Genetic transfer in plants through interspecific protoplast fusion. Science (Wash DC) 219:683–688

Shimamoto K, King PJ (1983) Isolation of a histidine auxotroph of *Hyoscyamus muticus* during attempts to apply BUdR enrichment. Mol Gen Genet 189:69–72

Sidorov VA, Maliga P (1982) Fusion-complementation analysis of auxotrophic and chlorophyll-deficient lines isolated in haploid *Nicotiana plumbaginifolia* protoplast cultures. Mol Gen Genet 186:328–332

Sidorov VA, Menczel L, Nagy F, Maliga P (1981) Chloroplast transfer in *Nicotiana* based on metabolic complementation between irradiated and iodoacetate treated protoplasts. Planta (Berl) 152:341–345

Skarzhynskaya MV, Cherep NN, Gleba Y (1982) Potato and tobacco hybrid cell lines and plants obtained by cloning individual protoplast fusion products. Sov Cytol Genet 6:42–48

Smillie RM, Melchers G, v. Wettstein D (1979) Chilling resistance of somatic hybrids of tomato and potato. Carlsberg Res Commun 44:127–132

Smith HH, Kao KN, Combatti NC (1976) Interspecific hybridization by protoplast fusion in *Nicotiana*. Heredity 67:123–128

Steffen A, Schieder O (1983) Selection and characterization of nitrate reductase deficient mutants of *Petunia*. In: Potrykus I, Harms CT, Hinnen A, Hütter R, King PJ, Shillito RD (eds) Protoplasts 1983 – poster proc. Birkhäuser, Basel, pp 162–163

Steffen A, Schieder O (1984) Biochemical and genetical characterization of nitrate reductase deficient mutants of *Petunia*. Plant Cell Rep 3:134–137

Stubbe W (1964) The role of the plastome in evolution of the genus *Oenothera*. Genetica 35:28–33

Stubbe W, Raven PH (1979) A genetic contribution to the taxonomy of *Oenothera* (including subsections *Euoenothera, Emersonia, Raimannia* and *Munzia*). Plant Syst Evol 133:39–59

Subrahmanyam NC, Kasha KJ (1973) Gene expression in haploid and hybrid progeny from crosses between *Hordeum vulgare* and *H. bulbosum*. Crop Sci 13:749–750

Syono K, Nagata T, Suzuki M, Kajita S, Mutsui C (1979) Fusion of pea root nodule protoplasts with tobacco mesophyll protoplasts. Z Pflanzenphysiol 95:449–457

Tabaeizadeh Z, Bergounioux C, Perennes C (1983) Increasing the variability of *Lycopersicon* Mill. by protoplast fusion with *Petunia* L. In: Potrykus I, Harms CT, Hinnen A, Hütter R, King PJ, Shillito RD (eds) Protoplasts 1983 – poster proc. Birkhäuser, Basel, pp 90–91

Téoulé E (1983) Somatic hybridization between *Medicago sativa* L. and *Medicago falcata* L. CR Acad Sci 297, S III:13–16

Thompson TE, Axtell JD (1978) Graft-induced transmission of cytoplasmic male sterility. J Hered 69:159–165

Townsend CHO (1897) Der Einfluß des Zellkerns auf die Bildung der Zellhaut. Jahrb Wiss Bot 30:484

Uchimiya H (1982) Somatic hybridization between male sterile *Nicotiana tabacum* and *N. glutinosa* through protoplast fusion. Theor Appl Genet 61:69–72

Uchimiya H, Ohgawara T, Kato H, Akijama T, Harada H, Sugiura M (1983) Detection of two different genomes in parasexual hybrids by ribosomal RNA genes analysis. Theor Appl Genet 64:117–118

Uchimiya H, Kobayashi S, Ono M, Brar DS, Harada H (1984) Characterization of nuclear and cytoplasmic information in the progeny of a somatic hybrid between male sterile *Nicotiana tabacum* and *N. glutinosa*. Theor Appl Genet 68:95–100

Vasil IK (1984) Cell culture and somatic cell genetics of plants. Vol 1. Laboratory procedures and their applications. Academic Press, New York

Vatsya B, Bhaskaran S (1981) Production of subprotoplasts in *Brassica oleracea* var *capitata* – a function of osmolarity of the media. Plant Sci Lett 23:277–282

Wallin A, Savage R (1982) A method for selection of heterokaryons from products of induced plant protoplast fusion. In: Fujiwara A (ed) Plant tissue culture 1982. Maruzen, Tokyo, pp 621–624

Wallin A, Glimelius K, Eriksson T (1974) The induction of aggregation and fusion of *Daucus carota* protoplasts by polyethylenglycol. Z Pflanzenphysiol 74:64–80

Wallin A, Glimelius K, Eriksson T (1978) Enucleation of plant protoplasts by cytochalasin B. Z Pflanzenphysiol 87:333–340

Wallin A, Glimelius K, Eriksson T (1979) Formation of hybrid cells by transfer of nuclei via fusion of miniprotoplasts from cell lines of nitrate reductase-deficient tobacco. Z Pflanzenphysiol 91:89–94

Weber G, Constabel F, Williams F, Fowke L, Gamborg OL (1976) Effect of preincubation of protoplasts on PEG-induced fusion of plant cells. Z Pflanzenphysiol 79:459–464

Wetter LR (1977) Isoenzyme patterns in soybean-*Nicotiana* somatic hybrid cell lines. Mol Gen Genet 150:231–235

Wetter LR, Kao KN (1976) The use of isoenzymes in distinguishing the sexual and somatic hybrids in callus cultures derived from *Nicotiana*. Z Pflanzenphysiol 80:455–462

Wetter LR, Kao KN (1980) Chromosome and isoenzyme studies on cells derived from protoplast fusion of *Nicotiana glauca* with *Glycine max* – *Nicotiana glauca* cell hybrids. Theor Appl Genet 57:273–276

Wettstein D, von, Poulsen C, Holder A (1978) Ribulose-1,5-bisphosphate carboxylase as a nuclear and chloroplast marker. Theor Appl Genet 53:193–198

White DWR, Vasil IK (1979) Use of amino acid analogue-resistant cell lines for selection of *Nicotiana sylvestris* somatic cell hybrids. Theor Appl Genet 55:107–112

Willecke K (1978) Review: Results and prospects of chromosomal gene transfer. Theor Appl Genet 52:97–104

Williamson FA, Fowke LC, Weber G, Constabel F, Gamborg OL (1977) Microfibril deposition on cultured protoplasts of *Vicia hajastana*. Protoplasma 91:213–320

Willis GE, Hartmann JX, de Lamater ED (1977) Electron microscopic study of plant – animal cell fusion. Protoplasma 91:1–14

Winkler H (1938) Über einen Burdonen von *Solanum lycopersicum* und *Solanum nigrum*. Planta (Berl) 27:680–707

Withers LA, Cocking EC (1972) Fine structural studies on spontaneous and induced fusion of higher plant protoplasts. J Cell Sci 11:59–75

Wright WE (1978) The isolation of heterokaryons and hybrids by a selective system using irreversible biochemical inhibitors. Exp Cell Res 112:395–407

Wullems GJ, Molendijk L, Schilperoort RA (1979) The expression of tumor markers in intraspecific somatic hybrids of normal and crown gall cells from *Nicotiana tabacum*. Theor Appl Genet 56:203–208

Xuan LT, Grafe R, Müller AJ (1983) Complementation of nitrate reductase deficient mutants in somatic hybrids between *Nicotiana* species. In: Potrykus I, Harms CT, Hinnen A, Hütter R, King PJ, Shillito RD (eds) Protoplasts 1983 – poster proc. Birkhäuser, Basel, pp 124–125

Zelcer A, Aviv D, Galun E (1978) Interspecific transfer of cytoplasmic male sterility by fusion between protoplasts of normal *Nicotiana sylvestris* and X-ray irradiated protoplasts of male-sterile *N. tabacum*. Z Pflanzenphysiol 90:397–407

Zenkteler M, Melchers G (1978) In vitro hybridization by sexual methods and by fusion of somatic protoplasts. Theor Appl Genet 52:81–90

Zimmermann U, Scheurich P (1981) High frequency fusion of plant protoplasts by electric fields. Planta (Berl) 151:26–32

Addendum to Chapters II and III

Only a few of a number of recent publications on somatic cell hybridization can be mentioned here. Fowke and Constabel (4) edited a book on protoplast technology with special reference to applied aspects. – *Fusion induction* by PEG, Ca^{2+} and high pH could be improved (7). – *Single fusion bodies* were cultured after mechanic isolation (5) and fusing single pairs of protoplasts (8). – *The fates of organelles* and their DNA were investigated (3, 9, 10, 11) – *Non-random plastid segregation* was indicated (10). – *Incompatibility* of *Nicotiana* and *Petunia* was confirmed (13; cf. p. 64). – *Economic species* of *Solanum* (2, 6), *Lycopersicon* (1, 5, 6, 11) and crucifers (3) received increased attention. – *Citrus* (x) *Poncirus* hybrids regenerated via embryogenesis.

References: 1. Adams TL, Quiros CF (1985) Plant Sci 40:209–219 – 2. Austin S, Ehlenfeld MK, Maer MA, Helgeson YP (1986) Theor Appl Genet 71:682–690 – 3. Chetrit P, Mathieu C, Vedel F, Pelletier G, Primard C (1985) Theor Appl Genet 69:361–366 – 4. Fowke LC, Constabel F (1985) Plant Protoplasts, CRC Press, 1985 – 5. Hamill YD, Pental D, Cocking EC (1985) Theor Appl Genet 71:486–490 – 6. Handley LW, Nickels RL, Cameron MW, Moore PP, Sink KC (1985) Theor Appl Genet 71:691–697 – 7. Kao KN, Saleem M (1986) J. Plant Physiol 122:217–225 – 8. Koop H-U, Schweiger H-G (1985) Eur J Cell Biol 39:46–49 – 9. Medgyesy P, Golling R, Nagy F (1985) Theor Appl Genet 70:590–594 – 10. Müller-Gensert E, Schieder O (1985) Curr Genet 10: 335–337 – 11. O'Connell MA, Hanson MR (1985) Theor Appl Genet 70:1–12 – 12. Ohgawara T, Kobayashi S, Ohgawara E, Uchimiya H, Ishii S (1985) Theor Appl Genet 71:1–4 – 13. Pental D, Hamill YD, Cocking EC (1986) Molec Gen Genet 202:342–347.

IV Molecular Biology of Plant Cell Transformation

N. S. YADAV [1]

1 Introduction

Recent advances in plant tissue culture techniques, recombinant DNA technology, and bacterial genetics, have made it feasible to isolate specific genes, manipulate them in vitro, and introduce them into plant cells. This not only opens up the exciting possibility of genetically manipulating crop plants, but also provides a powerful tool for studying regulation of plant gene expression and, possibly, for molecular cloning of selectable plant genes from gene libraries. Transformation, in this review, refers to the stable introduction of foreign genetic material into cells. There are several potential plant transformation vectors, DNA vehicles required for the efficient introduction and replication of foreign genes in cells (reviewed by Howell 1982). However, only the Ti-plasmid, the casual agent of crown gall tumorigenesis and, to a much more limited extent, the *ca*uliflower *m*osaic *v*irus (CaMV) have been successfully used to propagate foreign sequences in plants.

This chapter reviews mainly the molecular biology of crown gall tumorigenesis, since the rapid progress made in it in the past year or two has been, and will continue to be, important in the development of novel Ti-plasmid based transformation vectors. Moreover, Ti-plasmid genes serve as a useful model for the integration and expression of foreign genes in plants. The chapter also reviews the molecular biology of CaMV because recent progress made in it illustrates the potential use of plant viruses as alternate transformation vectors. There have been several recent reviews on Ti-plasmids (Bevan and Chilton 1982b; Ream and Gordon 1982; Howell 1982; Kemp 1983; Depicker et al. 1983; Caplan et al. 1983; Hooykaas and Schilperoort 1984; Binns 1984), including a book (Kahl and Schell 1982), and on CaMV (Howell 1982; Hohn et al. 1982; Hohn and Hohn 1982; Gardner 1983). This review covers the recent advances in the molecular biology of Ti-plasmid and CaMV that pertain to their present and potential use as plant gene vectors.

[1] E. I. du Pont de Nemours & Company, Central Research and Development Department, Wilmington, Delaware 19898, U.S.A.

Results and Problems in Cell Differentiation 12
Differentiation of Protoplasts and of Transformed Plant Cells (Edited by J. Reinert and H. Binding)
© Springer-Verlag Berlin Heidelberg 1986

2 Ti-Plasmids as Natural Plant Transformation Vectors

2.1 Crown Gall and Hairy Root Diseases: An Overview

The crown gall and hairy root diseases are neoplasms of several dicotyledonous plants caused by virulent strains of *Agrobacterium tumefaciens* and *A. rhizogenes,* respectively, Gram-negative soil bacteria that belong to the family Rhizobiaceae. Crown gall is a tumorous outgrowth with little or no differentiation at the site of infection. Hairy root is characterized by the proliferation of adventitious roots at the site of infection. Both neoplasms have two major characteristics that distinguish them from normal plant cells. First, they are stably altered in their growth phenotype: crown gall cells can grow axenically in vitro in the absence of exogenous plant growth hormones, auxin and cytokinin, and are unable to regenerate plants, while hairy roots differ from normal roots in their morphology and rapid growth rate. Second, both tissues synthesize novel low molecular weight metabolites, generically called opines, which can serve as specific nutrients for the inciting strain of *Agrobacterium.*

Virulence of *Agrobacteria* is associated with a class of large plasmids (180–240 kb) called Ti-(*t*umor *i*nducing) (Van Larebeke et al. 1974; Watson et al. 1975) or Ri-(*r*oot *i*nducing) (White and Nester 1980a) plasmids in *A. tumefaciens* or *A. rhizogenes,* respectively. Classification of *Agrobacterium* species on the basis of phytopathogenicity is superficial because, for example, the transfer of the Ri-plasmid from a virulent strain of *A. rhizogenes* to an avirulent strain of *A. tumefaciens* results in a virulent strain of *A. tumefaciens* that can now incite hairy roots (Albinger and Beiderbeck 1977). A segment of the Ti- and Ri-plasmids, called the T-(*t*ransferred) region, is transferred to and stably maintained in the transformed plant cells, where it is called T-DNA (Chilton et al. 1977; Chilton et al. 1982; White et al. 1982; Willmitzer et al. 1980, 1982a). The T-DNA is covalently attached to the host nuclear DNA (Yadav et al. 1980; Thomashow et al. 1980b; Zambryski et al. 1980) and, as detailed below, the expression of specific T-DNA genes confers the neoplasm phenotypes on the tranformed cells.

The Ti- and Ri-plasmids not only dictate the set of opines (see below) that are synthesized in the neoplasms, but also confer on the bacteria harboring them the ability to utilize the same set of opines as a source of nitrogen and carbon. The opines can also induce conjugational transfer of Ti-plasmids among *Agrobacteria* (see Kerr and Ellis 1982). Clearly, Ti- and Ri-plasmids are catabolic plasmids which have evolved a unique strategy of parasitism in higher plants through genetic transformation of plant cells.

2.2 The Ti- and Ri-Plasmids

There is great diversity both between and among the Ti- and Ri-plasmids. Opine metabolism, a plasmid trait which is linked to pathogenicity,has been used to classify the Ti- and Ri-plasmids into four and two major types, respectively (Table 1). Recently, a fifth type of Ti-plasmid, the succinamopine type, closely related to the nopaline Ti-plasmids, has been proposed (Chilton et al. 1984). The

Table 1. Classification of Ti- and Ri-plasmids

Plasmid type	Plasmid compatibility group	Example	Opines synthesized in neoplasm
Octopine Ti-plasmids (wide host range)	Rh-1	pTiA6	Octopine and Agropine families
Octopine Ti-plasmids (narrow host range)	Rh-1	pTiAg57	Octopine family
Nopaline Ti-plasmids	Rh-1	pTiC58	Nopaline family, Agrocinopines A and B
Agropine Ti-plasmids	Rh-2	pTi542	Agropine family, Agrocinopines C and D
Agropine Ri-plasmids	Rh-3	pRi1855	Agropine family, Agrocinopines A and B
Mannopine Ri-plasmids	Rh-3	pRi8196	Agropine family (except agropine); Agrocinopines C and D

Data from Hooykaas and Schilperoort (1984), Petit et al. (1983), and Lahners et al. (1984).

biosynthesis of the octopine family of opines (condensation products between pyruvate and an amino acid) is catalyzed by octopine synthase (Otten et al. 1977). The nopaline family of opines (condensation products between α-ketoglutarate and an amino acid) is synthesized by nopaline synthase (Kemp 1982). The agropine family of opines consists of condensation products between mannose and an amino acid (Tate et al. 1982). Agrocinopines are phosphorylated sugar derivatives of unknown structures (Ellis and Murphy 1981).

The relatedness between the diverse Ti- and Ri-plasmids is reflected by their plasmid incompatibilities (Table 1), DNA sequence homologies, and genetic complementations. As a group the wide host octopine and nopaline Ti-plasmids share four regions of extensive DNA sequence homology (Engler et al. 1981). One of these regions of homology, called the "common DNA" is found in the T-region (Fig. 2), suggesting common tumorigenicity functions in these Ti-plasmids (Chilton et al. 1978). Another region of homology is located outside the T-region and overlaps the *vir*-region, a region essential for tumorigenicity (see Sect. 2.3). In contrast, the other groups of Ti- and Ri-plasmids share only limited homology with the wide host octopine and nopaline Ti-plasmids (see review by Hooykaas and Schilperoort 1984), chiefly in the *vir*-region (White and Nester 1980b; Thomashow et al. 1981; Risuleo et al. 1982). The Ri-plasmids also hybridize weakly to other regions of the wide host octopine and nopaline Ti-plasmids, primarily in regions involved in opine metabolism (White and Nester 1980b; Willmitzer et al. 1982a; Lahners et al. 1984). The *vir*-region of the wide host octopine Ti-plasmid can be complemented in trans by the narrow host range octopine and nopaline Ti-plasmids and by Ri-plasmids (Hoekema et al. 1984). These data suggest that there are common virulence functions in these diverse plasmids. It is evident from Table 1 that different types of Ti- and Ri-plasmids synthesize different combinations of families of opines. In fact, the Ri-plasmids may be considered a subset of Ti-plasmids. Such a view is supported by the finding that some mutant Ti-plasmids incite tumors with roots (see Sect. 2.6.3.2) and Ri-plasmids incite tumors and not roots on some host tissues (White et al. 1982). These different plasmids provide a good opportunity to study plasmid evolution.

The following discussion is restricted to the molecular biology of the wide host octopine and nopaline Ti-plasmids, which have been much better characterized that the other Ti- and Ri-plasmids. Tumors incited by nopaline and octopine Ti-plasmids will be referred to as nopaline and octopine tumors, respectively.

2.3 The *Vir*-Region and the Early Stages of Transformation

The powerful technique of transposon mutagenesis has been successfully exploited in the genetic mapping of Ti-plasmid functions involved in crown gall tumorigenesis. Because transposons are genetically marked with selectable genes, their insertion can be conveniently followed. Furthermore, because of their large size they can be easily located on the physical map of the target sequence by restriction endonuclease analysis. By screening thousands of Ti-plasmids carrying apparently random transposon insertions, several mutant plasmids were identified that were altered in their ability to form tumors (Garfinkel and Nester 1980; Ooms et al. 1980; Holsters et al. 1980). Some of these Ti-plasmid mutations map in the T-region which is transferred to plants. However, most map in a region of 30–50 kb called the *vir*-region which is located to the left of the T-region on the Ti-plasmids and which is not maintained in plant cells. Few avirulence mutations mapped near the replicator region of the Ti-plasmid (Koekman et al. 1982) and some of these were subsequently shown not to be involved in virulence per se (Hille et al. 1982). Remarkably, all avirulence insertion mutations mapped outside the T-region, whereas most mutations affecting the morphology and size of the tumor mapped to the T-region (discussed in Sect. 2.6.3). However, few tumor morphology mutations also mapped to the *vir*-region (Garfinkel and Nester 1980).

Detailed genetic analysis of the *vir*-region of octopine and nopaline Ti-plasmids has identified, respectively, eleven and six complementation groups of genetic loci that affect virulence (Iyer et al. 1982; Klee et al. 1982; Hille et al. 1984; Lundquist et al. 1984). All *vir*-region mutations tested were complemented in trans by the wild type *vir*-region when present on an R prime plasmid (Hille et al. 1984). The *vir*-region can function in trans with respect to the T-region, since *Agrobacteria* containing the *vir*- and T-regions on different plasmids are virulent (Hoekema et al. 1984). These results suggest that the *vir*-region is expressed in the bacterium. Some virulence loci also map to the bacterial chromosome (Garfinkel and Nester 1980). No function is known for any of the virulence loci; they could be involved in the early events of transformation, for example, in the transfer of the T-region and in determining the host range (Liu et al. 1982).

Our knowledge of the early stages of tumor induction is scant. We know that wounding is essential for tumor formation, possibly for providing entry to the bacteria into the intercellular space and for "conditioning" the host cell (Braun 1982). *Agrobacteria* do not enter the host cell, but attach to specific sites on the host cell wall (reviewed by Hookyaas and Schilperoort 1984; Matthysee 1983).

There is evidence that this binding involves bacterial lipopolysaccharides (Banerjee et al. 1981) and cellulosic microfibrils (Matthysee 1983) as well as plant pectic material (Rao et al. 1982). Plant cells in the wound are susceptible to infection after about 24 to 48 h following wounding ("inception phase") and remain susceptible for some time before becoming refractory to infection (Braun 1952). It has been proposed that cell division is required for tumor initiation since the inception phase correlates with the time required for wound-induced cell division (Braun 1982). Even *Agrobacterium* appears to need conditioning for about 10 h in the wound site before it is infective (Lipetz 1966). Transformation takes about 10 h (Lipetz 1966), requires live bacteria, and is sensitive to temperatures above 30 °C, although both bacteria and plants grow normally at these temperatures (Braun 1952). Once induced, however, tumor formation is unaffected by the absence of the bacteria or by higher temperatures. Since conjugal transfer of Ti-plasmids between different strains of *A. tumefaciens* is similarly thermosensitive, it was suggested that bacterial conjugation and plant transformation share a thermosensitive step (Tempe et al. 1977). However, all bacterial mutants unable to conjugate remain virulent (Klapwijk et al. 1978).

2.4 T-DNA Structure

The structure of T-DNA has been analyzed in detail by Southern blot hybridization in about a half dozen nopaline (Lemmers et al. 1980; Hepburn et al. 1983 a) and two dozen octopine (Thomashow et al. 1980 a; De Beuckleer et al. 1981; Ooms et al. 1982 a; Urisic et al. 1983) tumor cell lines, mostly of tobacco. These studies revealed that the tumor DNA contains fragments that hybridize only to the T-region of the Ti-plasmid. The T-DNAs in octopine and nopaline tumors are present as contiguous segments of 13 kb and 23 kb, respectively, of the Ti-plasmids. Both of these T-DNAs contain the "common" DNA conserved between wide host octopine and nopaline Ti-plasmids (Chilton et al. 1978) (Fig. 2). In octopine tumors this T-DNA is called T_L-DNA, because in some lines an additional 6 kb segment of the octopine Ti-plasmid, called T_R, is present. The T_L-region is present to the left of the T_R-region on the Ti-plasmid (Fig. 2). The copy number and the sites of integration of T_R-DNA, if present in the tumor, are usually independent of those of T_L DNA. The T-DNA borders in different tumors are apparently fixed. For instance, the T-DNA borders were unaltered in a tumor incited by a Ti-plasmid carrying an insertion of a 15 kb foreign DNA (transposon Tn7) in the T-region (Lemmers et al. 1980).

Fragments containing tumor DNA which do not comigrate with any T-region fragment and which hybridize to probes from either the left or the right ends of the T-regions presumably are "border" fragments that contain T-DNA and plant DNA junctions. The size of the border fragments in different tumor lines is different (Lemmers et al. 1980; Thomashow et al. 1980 a; De Beuckleer et al. 1981; Ooms et al. 1982 a; Hepburn et al. 1983 a; Urisic et al. 1980). For example, 11 independent tumor lines derived from isogenic sunflower plants had 11 distinct T-DNA insertion patterns (Urisic et al. 1983). Thus, the site of T-DNA integration, while precise with respect to the Ti-plasmid, is apparently random with respect to the host DNA. It has yet to be determined if there is host sequence specificity for T-DNA integration. For this, DNA sequences of several T-DNA/plant DNA border fragments will have to be compared with that of the corresponding region from normal host plants.

Some nopaline (Lemmers et al. 1980; Zambryski et al. 1982) and octopine (Ooms et al. 1982a; Holsters et al. 1983) tumors have DNA fragments that hybridize to fragments from both ends of the T-region. The presence of these fragments, called "fusion fragments", suggests, in the absence of any evidence for circular T-DNA (Lemmers et al. 1980), tandem T-DNA sequences in these tumor cell lines.

However, even within the limits of resolution of these studies, the ends of T-DNA are not invariant (Thomashow et al. 1980a; De Beuckleer et al. 1981). Abnormal T-DNA insertions, often severely deleted, have been observed in some tumor cell lines (Ooms et al. 1982a; Hepburn et al. 1983a). Hepburn et al. (1983a) reported three kinds of abnormal truncated T-DNA insertions: insertions having normal left and abnormal right borders, insertions having normal right and an abnormal left borders, and insertions having homology to several noncontiguous regions of T-DNA. It is not known whether these abnormalities, or the minor variations in the ends of T-DNA, result from primary events utilizing alternate sites for T-DNA integration (see Sect. 2.5) or represent deletions or rearrangements of T-DNA in tissue culture subsequent to its normal integration.

Southern blot analysis has also been used to estimate the copy number of T-DNA in tumor cells, which varies from a single copy to several copies. However, these estimates are complicated by the use of uncloned tumor cell lines or by the use of cloned lines that have been in culture for several years. Usually multiple border fragments, representing more than one T-DNA insertion, observed in uncloned primary tumor cells, can be resolved by cellular cloning (Ooms et al. 1982a).

For more detailed analysis of the T-DNA structure, T-DNA fragments have been isolated by molecular cloning from different nopaline (Zambryski et al. 1980; Yadav et al. 1980; Zambryski et al. 1982) and octopine (Holsters et al. 1982; Thomashaw et al. 1980b; Holsters et al. 1983; Urisic et al. 1983) tumor cell lines. Some isolated border fragments hybridize to unique DNA of normal plants, but most hybridize to repetitive sequences in normal plants. Restriction endonuclease mapping of these clones confirmed that, except in one case, the T-DNA had not undergone any detectable rearrangement after several years (23 years in one case) in plant tissue culture. The only exception was an octopine tumor line where a duplication of a 0.5 kb region (from the middle of the T-DNA) occurred at the T-DNA/plant DNA junction (Thomashow et al. 1980b; Simpson et al. 1982).

The nucleotide sequences around the putative ends of the T-DNA in these isolated border and fusion fragments have been compared with those in the corresponding regions of the Ti-plasmids (see below) (Zambryski et al. 1980; Yadav et al. 1982; Zambryski et al. 1982; Simpson et al. 1982; Holsters et al. 1983). From these comparisons it became evident that the T-DNA from the tumor and the T-region from the Ti-plasmid are colinear at the molecular level too, and the ends of T-DNA (the putative junctions) were identified as the points where the two sequences begin to diverge. However, the nucleotide sequence beyond the point of divergence in one T-DNA border fragment suggests that it is not of plant origin, but results from rearrangements of Ti-plasmid end sequences (Simpson et al. 1982). In addition, the nucleotide sequences at the junction of the left and the right T-DNA ends in the T-DNA fusion fragments isolated from tumors show that they are not exact fusions, instead they contain variable sequences of uncertain origin at the junction (Zambryski et al. 1980, 1982; Holsters et al. 1983). In one case the junction sequence appears to be rearranged Ti-plasmid end sequences (Zambryski et al. 1980) and in another it appears to have repeated plant sequences (Holsters et al. 1983).

2.5 T-DNA Border Repeats and Their Role in T-DNA Transfer

The ends of the T-region of a nopaline Ti-plasmid (Yadav et al. 1982; Zambryski et al. 1982) as well as the T_L- and T_R-regions of octopine Ti-plasmids (Holsters et al. 1983; Barker et al. 1983; Gielen et al. 1984) have been sequenced. The most striking feature exhibited at the ends of all T-regions is a 24 bp direct, imperfectly repeated sequence (18–20/24 bases matching perfectly) (Fig. 1). No larger homology was detected either from the nucleotide sequences or by Southern blot hybridization (Yadav et al. 1981). Interestingly, a *chi* sequence, 5′-GCCTGGTGG-3′, a sequence involved in bacterial recombination (Smith 1983) is present adjacent to the left repeat of the nopaline T-region (Yadav et al. 1982).

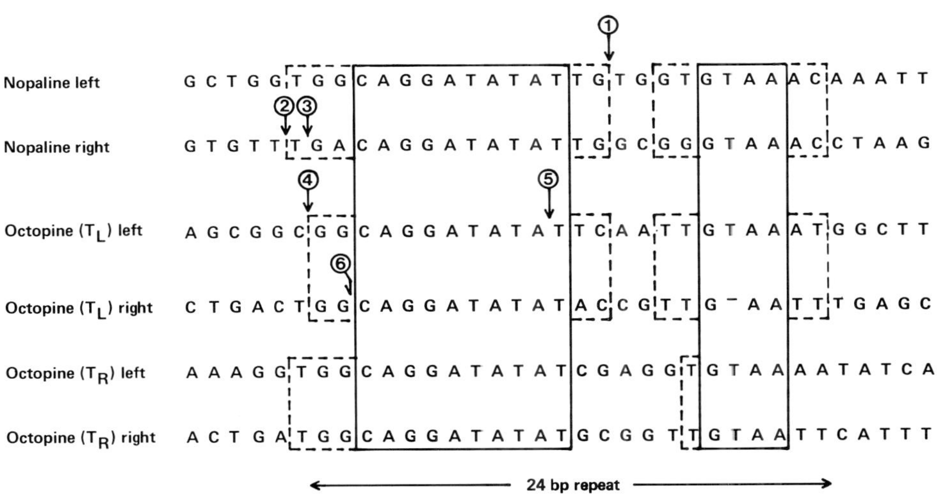

Fig. 1. Border sequences of T-regions on nopaline and octopine Ti-plasmids The sequences have been aligned to reveal two boxes of perfect sequence homology between the different 24 bp direct repeat sequences (see text). The homology between the left and right borders of a T-region can be extended as shown by the *stippled boxes*. *Positions 1* to *6* are points where these sequences diverge from the corresponding T-DNA ends in different T-DNA border and fusion fragments isolated from tumors. T_L and T_R are the two T-regions on octopine Ti-plasmids (see Fig. 2). Data taken from Zambryski et al. (1982), Yadav et al. (1982), Simpson et al. (1982), Holsters et al. (1983), and Barker et al. (1983)

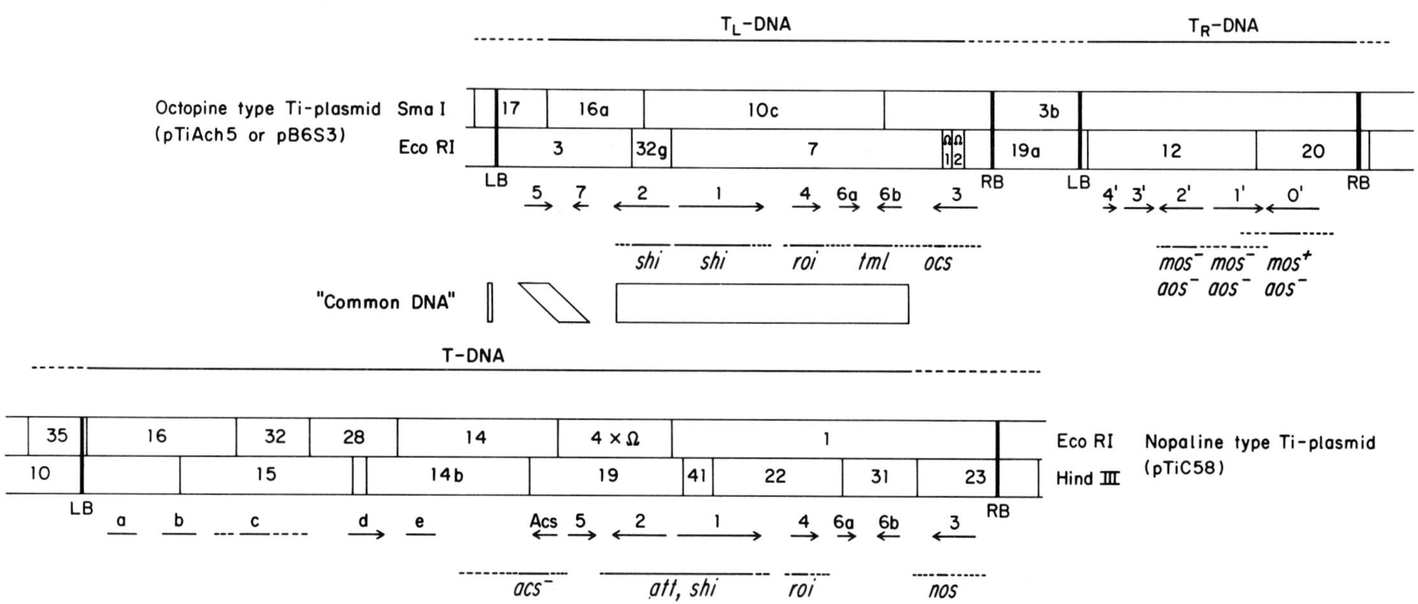

Fig. 2. Physical, transcriptional, and genetic maps of the T-regions of octopine and nopaline Ti-plasmids. The *left* (*LB*) and *right* (*RB*) borders (see Fig. 1) of each T-region are shown as *thick vertical bars* across the physical maps. The extent of the T-DNA's in most tumors is shown above the physical maps, with the *stippled regions* defining the limits of T-DNA borders determined by Southern analysis (see text). The location and polarity (if known) of T-DNA transcripts in tumors is shown below the physical maps and above the genetic maps. The "common DNA" represents the sequence homology between octopine and nopaline Ti-plasmids. *Shi, roi, tml, att* are genetic loci for tumor morphology mutations affecting shoot inhibition, root inhibiting, large tumor size, and tumor attenuation, respectively; *ocs, nos, acs, mos,* and *aos* are genetic loci affecting synthesis of octopine, nopaline, agrocinopine, mannopine, and agropine, respectively. Data compiled from Garfinkel et al. (1981), Bevan and Chilton (1982b) Barker et al. (1983), Velten et al. (1983), Willmitzer et al. (1983), Joos et al. (1983a), Gielen et al. (1984), and Salomon et al. (1984)

However, the significance of this sequence is questionable since it is not found around the repeats in octopine Ti-plasmids.

The significance of the short, direct repeats flanking the T-region of the Ti-plasmid becomes obvious by comparing them with the endpoints of T-DNA in the T-DNA border and fusion fragments isolated from tumors (Zambryski et al. 1980; Yadav et al. 1982; Zambryski et al. 1982; Simpson et al. 1982; Holsters et al. 1983). Of the seven sequenced left ends of nopaline and octopine T-DNAs, three extended up to the left repeat sequence (to position numbers 1, 4, and 5 in Fig. 1) and four were 57 bp, 86 bp, 92 bp, or 93 bp short of the left repeat sequence (not shown in Fig. 2). Of the six sequenced right ends of nopaline and octopine T-DNAs, five extended to the right repeat sequence (three to position number 2 and one each to position numbers 3 and 6 in Fig. 1) and one was 7 bp short of the repeat sequence (not shown in Fig. 2).

While there is strong circumstantial evidence for the involvement of both 24 bp repeat sequences in the excision/integration of the T-region, genetic evidence for the role of the left repeat sequence is less convincing. Whereas Ti-plasmids deleted for the left border of the T-region are able to incite normal tumors (Leemans et al. 1982; Joos et al. 1983 a, b), those deleted for the right border of the T-region were only weakly virulent (Leemans et al. 1982; Ooms et al. 1982 a; Joos et al. 1983 a; Hille et al. 1983). In the case of octopine Ti-plasmids this weak virulence was observed only when the right borders of both the T_L- and T_R-regions were deleted. This result suggests that the right border of either the T_L- or T_R-region is required for normal tumor formation. In the latter case the T_L- and T_R-DNA would be transferred as one contiguous segment, as is found in some natural octopine tumors (Urisic et al. 1983). This interpretation is supported by the observation that tumors incited by octopine Ti-plasmids lacking only the right border of the T_L-region produced agropine, a marker for the T_R-DNA (Hille et al. 1983). In other experiments, *A. tumefaciens* harboring two plasmids, a Ti-plasmid deleted for most or all of the T-region, and a second plasmid containing only the T-region, was found to be virulent only if the T-region on the second replicon was flanked by its border sequences (deFramond et al. 1983; Joos et al. 1983 b). Taken together, these genetic experiments indicate that at least the right T-region border sequence, which contains the 24 bp repeat, is important for normal tumor formation. The right T-region borders have been proposed to provide active function(s) in cis that is (are) involved in the transfer of T-DNA to plant cells (Joos et al. 1983 b). Apparently, the left 24 bp repeat sequences are preferred T-region borders in whose absence secondary borders can be substituted. The nucleotide sequence of the T_L-region revealed ten sequences that resembled the consensus border repeat sequences, and which were suggested to be involved in the formation of truncated T-DNAs found in several tumors (Gielen et al. 1984).

The mechanism by which T-DNA is transferred into the genome of the plant cell is not known. For example, we do not know if T-DNA excision from the Ti-plasmid involves bacterial and/or plant functions. There is conflicting evidence as to whether most (or all) of the Ti-plasmid enters the host plant cell or only the T-DNA alone enters the cell, following its excision from the Ti-plasmid. Tumor cells obtained from in vitro transformations of plant protoplasts with Ti-plasmid DNA (Krens et al. 1982; Draper et al. 1982), in contrast to those obtained from infection of protoplasts by virulent *A. tumefaciens* (Ooms et al. 1982 a; Fraley and Horsch 1983), lack normal T-DNA borders. These observations suggest that bacterial functions are involved in the recognition of the normal T-DNA borders. However, Joos et al. (1983 b) found that most, if not all, of the Ti-plasmid enters the plant cell. An avirulent deletion mutant of a Ti-plasmid was made by removing most of its T-region, including the genes essential for tumorigenicity (see Sect. 2.6.3), but which retained the T-region borders. Reinsertion of a borderless T-region outside the deleted T-region of this mutant Ti-plasmid resulted in a modified Ti-plasmid that could incite plant tumors, albeit weakly. It remains to be seen if the transfer of most of the Ti-plasmid into host cells represents a normal feature of transformation.

T-DNA is a unique mobile genetic element since it lacks several features of known prokaryotic and eukaryotic transposons (reviewed by Calos and Miller 1980; Temin 1980), including a maize controlling element (Sutton et al. 1984). Unlike these transposons, which are characterized by inverted terminal repeats that are maintained intact during transposition, T-DNA has short, direct repeats that are apparently not preserved following transfer (Fig. 2). It is not known if T-DNA insertion into plant DNA results in host sequence duplication, another characteristic feature of transposons. To determine this, both ends of T-DNA will have to be isolated in one piece and its border sequences compared to that of the insertion site isolated from normal plant cells. The T-region on the Ti-plasmid does have some structural resemblance to the prophage of bacteriophage λ which is also flanked by short (15 bp), direct, imperfect repeats (Landy and Ross 1977). In analogy to the λ prophage, site-specific recombination between the T-region border repeats could result in the excision of a monomeric circular T-region carrying one copy of the repeat. The presence of tandem copies of T-DNA in some tumors is consistent with the presence, at some time, of circular T-DNA intermediates (Zambryski et al. 1980). Limited sequence homology can be detected between the T-DNA border repeat sequence and the putative plant context sequence in some T-DNA border fragments (Yadav et al. 1982; Zambryski et al. 1982). However, since these sequences are AT rich, these could be coincidences.

The discovery of the molecular mechanism of T-DNA transfer will be a major milestone in our understanding of the crown gall system as well as in designing new plant genetic engineering vectors. The development of new methods, such as protoplast cocultivation (Wullems et al. 1981 a; Fraley and Horsch 1983), should allow the capture and study of the molecular intermediates involved in the early stages of this process.

2.6 T-DNA Functions

The study of T-DNA function has involved construction of transcription and genetic maps of the T-DNA, and the identification of T-DNA encoded proteins. These studies show that the expression of specific T-DNA genes results in the tumor phenotypes: hormone-independent growth and opine biosynthesis.

2.6.1 T-DNA Transcription

T-DNA transcription studies identified up to eight T_L-DNA transcripts and five T_R-DNA transcripts in octopine tumors (Gelvin et al. 1982; Willmitzer et al. 1982b; Murai and Kemp 1982a; Willmitzer et al. 1983; Velten et al. 1983; Karcher et al. 1984; Winter et al. 1984), and up to thirteen distinct T-DNA transcripts in nopaline tumors (Bevan and Chilton 1982a; Willmitzer et al. 1983). The location and polarity (if determined) of these transcripts are shown in Fig. 2. The nucleotide sequences of the entire T_L-DNA (Barker et al. 1983; Gielen et al. 1984) and T_R-DNA (Barker et al. 1983) of an octopine Ti-plasmid have been published and open reading frames that correspond to the octopine T-DNA transcripts have been identified (Gielen et al. 1984; Barker et al. 1983).

Although T-DNA is of prokaryotic origin its transcripts have eukaryotic features. Polyadenylated T-DNA transcripts are found on polyribosomes (Willmitzer et al. 1981a; Schroder and Schroder 1982). In vitro translation of at least some of these transcriptions is inhibited by the cap analog pm7_G, suggesting that the mRNAs are capped (Schröder and Schröder 1982). T-DNA is transcribed by the α-amanitin sensitive RNA polymerase II (Willmitzer et al. 1981b). All five T_R-DNA and four T_L-DNA transcripts from octopine tumors (De Greve et al. 1983; Dhaese et al. 1983; Klee et al. 1984; Lichtenstein et al. 1984) and one T-DNA transcript from a nopaline tumor (Depicker et al. 1982; Bevan and Chilton 1982a) have been mapped on their genes by S1 nuclease protection. Sequences resembling the "TATA" box (consensus sequence TAT^A_TA^A_T; Breathnach and Chambon 1981) and in most cases, the "CCAAT" box (consensus sequence GG^C_TCAATCT; Benoist et al. 1980) are found 25–32 bp and 59–80 bp, respectively, upstream of the transcription initiation site. The TATA and CCAAT consensus sequences are present 25 bp and 80–105 bp, respectively, upstream of the transcription initiation sites of several eukaryotic genes and are believed to be involved in the accuracy and regulation of transcription initiation (Breathnach and Chambon 1981). Sequences resembling the consensus polyadenylation signal (AATAAA; Benoist et al. 1980) are also found within 50 bp of the 3′ end of the transcripts (see Dhaese et al. 1983). Similar consensus-like sequences have also been detected in the putative untranslated regions of other sequenced open reading frames to which transcripts have been assigned, but not mapped (Barker et al. 1983; Gielen et al. 1984). No consensus prokaryotic ribosome binding site (Rosenberg and Court 1979) is present upstream of the putative translation initiation codon in any of these open reading frames (Gielen et al. 1984; Barker et al. 1983). Furthermore, no general bias in codon usage has been found in these coding sequences (Depicker et al. 1982; Gielen et al. 1984; Lichtenstein et al. 1984). However, T-DNA genes do not appear to be interrupted by an intervening sequence (Depicker et al. 1982; Bevan et al. 1983a; Karcher et al. 1984; Winter et al. 1984), a feature of some animal and plant genes (see Gielen et al. 1984). In fact, all T_L-DNA genes appear to lack introns because the sizes of T_L-DNA transcripts correlate well with the sizes of both the open reading frames and the in vitro translation products (Schröder and Schröder 1982; Willmitzer et al. 1983; Gielen et al. 1984).

Transcription of T-DNA sequences does not appear to depend on promoters in the plant sequences that flank the T-DNA. Transcription of both T-DNA strands, the enormous variation in the relative abundance of the transcripts (Bevan and Chilton 1982a; Willmitzer et al. 1983; Gelvin et al. 1982) and the absence of polar effects of insertion mutations (Leemans et al. 1982; Joos et al. 1983a; Willmitzer et al. 1982b) suggest that most, if not all, T-DNA transcripts are initiated and terminated independently within the T-DNA. Deletion studies have demonstrated that the promoters for the nopaline synthase and octopine synthase genes, which are within 400 bp of the T-DNA borders, are contained within 261 bp and 295 bp, respectively, of their transcription initiation sites (Koncz et al. 1983).

The relative abundance of different T-DNA transcripts varies not only within a tumor tissue, but also between different tumors (Gelvin et al. 1982; Murai and Kemp 1982b; Willmitzer et al. 1983; Winter et al. 1984). Qualitative and quantitative differences in T-DNA transcription patterns have been correlated with the growth of tumors on solid medium or in suspension culture (Willmitzer et al. 1981a; Karchner et al. 1984), and with tumor morphology in culture (Willmitzer et al. 1983). The levels of opine, an easy marker to score, varied considerably between different subclones of a tumor line, during shoot regeneration following grafting, and in the progeny seeds (see Sect. 2.7) (van Slogteren et al. 1983; Barton et al. 1983). In one case studied, this variation in octopine level was also reflected in the level of octopine synthase mRNA (van Slogteren et al. 1983).

The mechanism of transcriptional modulation of T-DNA gene expression is not known, but it could, at least in some cases, involve DNA methylation. Using methylation-specific isoschizomers of restriction endonucleases, it was shown that while at least one copy of T-DNA in tumors was unmethylated at the sites tested, extensive methylation of most copies was detected when multiple copies of T-DNA were present (Gelvin et al. 1983; Hepburn et al. 1983a, b). Hepburn et al. (1983a, b) found that while a flax nopaline tumor line contained about 25 copies of the nopaline synthase gene (some unmethylated at the sites tested), there was no detectable nopaline synthase enzyme activity in tumor extracts. Growth of these cells in the presence of 30 μM 5-azacytidine, a demethylation agent, resulted in demethylation of an average of about two copies of the nopaline synthase gene per diploid cell and a concomitant severalfold increase in the levels of both nopaline synthase transcripts and the enzyme activity. These results have a strong bearing on plant genetic engineering efforts, that is, how to prevent suppression, by methylation, of foreign gene expression in the absence of a selection pressure.

A subset of six T-DNA transcripts are common to both octopine and nopaline tumors and map in the "common DNA" (Fig. 2). The corresponding transcripts 1, 2, 4, 5, 6a, and 6b in both types of tumors have the same size and polarity, similar map locations and, most importantly, nucleic acid sequence homology (Willmitzer et al. 1982b; Willmitzer et al. 1983). There is genetic evidence that the corresponding genes have homologous functions, too (see Sect. 2.6.3.2).

2.6.2 T-DNA-Encoded Proteins

In attempts to identify the T-DNA encoded proteins, T-DNA transcripts, isolated from octopine tumors by hybridization-selection to cloned T-region fragments, have been translated in vitro (McPherson et al. 1980; Schröder et al. 1981; Murai and Kemp 1982b; Schröder and Schröder 1982). Transcript 3 has been identified as the mRNA for octopine synthase by immunoprecipitation of its in vitro translation product by antiserum raised against the enzyme (Schröder et al. 1981; Murai and Kemp 1982b). Schröder and Schröder (1982) have identified two additional in vitro translation products of 14 kD and 27 kD of unknown function. On the basis of their size and hybridization selection, the 14 kD protein has been identified as a product of gene 7, while the 27 kD protein has been proposed to be a mixture of two proteins, one encoded by transcript 4 and the other by transcript 5 (Gielen et al. 1984). As the levels of the mRNAs for these proteins were at the detection limit of this technique, an attempt was made to identify the products of the other T_L-DNA genes by their expression in *E. coli* minicells. Four proteins were synthesized in *E. coli* from the T_L-region genes that map in the "common DNA" region (Schröder et al. 1983), three of which correspond closely in size and "map location" to the proteins predicted from the nucleotide sequence of the open reading frames assigned to transcripts 1, 2, and 4 (Fig. 2). The four proteins expressed in *E. coli* minicells were also expressed in an *A. tumefaciens* cell-free system (Schröder et al. 1983). These findings raise the interesting possibility that the conserved region of the T-DNA genes are expressed both in *A. tumefaciens* and in plant cells. In fact, Schröder et al. (1984) have demonstrated that the gene for transcript 2 is involved in the synthesis of indole acetic acid both in *Agrobacterium* and transformed plant cells. The expression of this gene both in *Agrobacterium* and plant cells may have significance during the early stages of transformation as well as in determining the host range (Liu et al. 1982).

2.6.3 Genetic Analysis of T-DNA Functions

2.6.3.1 Opine Biosynthesis

Transposon insertions, deletions, and site-specific mutagenesis have identified one T-DNA locus each for octopine synthase (Koekman et al. 1979; DeGreve et al. 1981; Garfinkel et al. 1981), nopaline synthase (Holsters et al. 1980; Joos et al. 1983a), and the biosynthesis of agricinopine (Joos et al. 1983a) (Fig. 2). Similarly, transposon insertions into the genes for each of the five transcripts of T_R-DNA, identified three loci involved in the production of mannopine and agropine in tumors (Salomon et al. 1984) (Fig. 2). This genetic evidence is consistent with biochemical data that mannopine is a precursor of agropine (Tate et al. 1982). Interestingly, this region has homology to the Ri-plasmids (Willmitzer et al. 1982a) which induce hairy roots containing mannopine and agropine. Genetic studies have also shown that opine production is not required for tumorigenicity and vice versa. The T-DNA genes for which no function is known (see below) might be involved in the biosynthesis of opines, which have not as yet been discovered.

2.6.3.2 Tumor Morphology

The virulence functions of the Ti-plasmids have been mapped by deletion and transposon mutagenesis (Garfinkel and Nester 1980; Holsters et al. 1980; De-

Greve et al. 1981; Ooms et al. 1981; Hille et al. 1982). While deletions of certain parts of the T-region result in loss of virulence, none of approx. 100 transposon insertions made in the T-region have caused avirulence (Garfinkel and Nester 1980; Ooms et al. 1981; Garfinkel et al. 1981; Joos et al. 1983a). However, some transposon insertions in the T-region resulted in tumors with altered morphology. Extensive mutagenesis by these transposon insertions as well as by site-specific insertions and small deletions (Leemans et al. 1981, 1982; Joos et al. 1983a; Ream et al. 1983; Barton et al. 1983; Hille et al. 1983) has defined up to three distinct genetic loci in octopine T_L-DNA and nopaline T-DNA that affect the morphology and size of tumors (Fig. 2). The *Shi* (*Sh*oot *i*nhibiting) and *Roi* (*ro*ot *i*nhibiting) loci in Fig. 2 correspond to the *Tms* (*t*umor *m*orphology *s*hooty) and *Tmr* (*t*umor *m*orphology *r*ooty) loci of Garfinkel et al. (1981). The *Tml* (*t*umor *m*orphology *l*arge) locus is expressed only in octopine tumors (Garfinkel et al. 1981; Joos et al. 1983a). While the wild-type Ti-plasmids incite undifferentiated tumors, Ti-plasmids with *shi* or *roi* mutation incite tumors in *Nicotiana tabacum* stems that form shoots or roots, respectively. Octopine Ti-plasmids with a *tml* mutation incite tumors that are two- to threefold larger in some plants, such as *Kalanchoe,* than those incited by the wild-type Ti-plasmids (Garfinkel et al. 1981). It is important to emphasize that the size and morphology of tumors result from a complex, poorly understood interaction between the inciting strain of bacterium and the physiological state of the host plant tissue. Thus, the same strain of bacterium can evoke different tumor responses in different plant species or even in different tissues of the same plant (De Cleene and De Ley 1981). Therefore, the comparison of tumor morphology phenotypes induced by these mutant Ti-plasmids is made on "indicator" plant tissues.

Interestingly, all genetic loci can be assigned to one or possibly two specific transcripts that map to the T-DNA region common to wide host octopine, nopaline, and agropine Ti-plasmids (Chilton et al. 1978) ("common DNA" in Fig. 2). Therefore, the genes (loci) for these transcripts will be referred to by the transcript number. The original *shi* locus which overlapped genes 1 and 2 can be split into two distinct loci since transposon insertions in the Eco R1 restriction site between the genes 1 and 2 (Fig. 1) are phenotypically silent (Klee et al. 1984). It is unclear if the *tml* is composed of gene 6a, gene 6b, or both.

The phenotypes of the tumor morphology genes, assuming no polar effects of their mutations, can be summarized in terms of specific T-DNA genes: genes 1 and 2 are both essential to inhibit shoot formation and gene 4 is essential to inhibit root formation (Leemans et al. 1981, 1982; Garfinkel et al. 1981; Joos et al. 1983a; Ream et al. 1983; Barton et al. 1983; Hille et al. 1983). However, the removal of gene 1 or 2 *and* gene 4 does not result in the formation of either shoots or roots at the site of infection: such double mutants are avirulent (although a weak tumor response is observed in some hosts which may be due to Ti-plasmid genes outside the T-region) (Ream et al. 1983; Hille et al. 1983; Inze et al. 1984). Thus, shoot formation in tumors requires both inactivation of genes 1 or 2 and expression of gene 4, while root formation in tumors requires both inactivation of gene 4 and expression of genes 1 and 2. Genes 6a and 6b together and gene 5 appear to be involved in determining the size of octopine and nopaline tumors, respectively, in some plants (Garfinkel et al. 1981; Leemans et al. 1982; Joos et al. 1983 a). Since *shi, tml* double mutants produce large galls without shoots or roots (Ream et al. 1983), genes 6a and 6b might act by either enhancing the effect of gene 4 or reducing the effect of genes 1 or 2.

There are two important conclusions from the genetic analysis described above: (1) a set of at least three T-DNA genes is involved in tumorigenicity, and (2) no T-DNA gene is required for the transfer, integration, and maintenance of T-DNA in transformed cells. This allowed the construction of an avirulent Ti-plasmid that was deleted for the entire T-region except for its borders and the nopaline synthase gene, but retained its ability to transform plant cells (Zambryski et al. 1983). This "disarmed" Ti-plasmid has been used to extend the genetic analysis of the tumor morphology genes by introducing them singly into plants and studying their effect on the transformed cells. While genes 1, 2, and 5 singly and genes 6a and 6b together were unable to form tumors, gene 4 alone induced tumors that formed nopaline positive shoots in vitro (Inze et al. 1984). In another approach, wounded tobacco stems were coinfected with pairs of *A. tumefaciens* containing Ti-plasmids with different tumor morphology mutations. Coinfection with a pair of Ti-plasmids, one carrying a *shi* mutation and the other a *roi* mutation, resulted in wild-type tumors (Ooms et al. 1981). Six cellular clones derived from one such tumor were shown to contain T-DNAs from each mutant Ti-plasmid, suggesting intracellular genetic complementation (Ooms et al. 1982 b). Coinfection with a pair of disarmed Ti-plasmids carrying only the nopaline synthase gene and either gene 1 or gene 2 resulted in small tumors that formed nopaline positive roots in vitro (Inze et al. 1984).

2.6.4 Role of Phytohormones in Tumorigenicity

An insight into the function of some tumor morphology loci has been obtained by complementation of mutations through the exogenous application of plant growth hormones (Ooms et al. 1981). Application of napthalene acetic acid (NAA), an auxin-like synthetic growth regulator, or cytokinin to tumors incited by Ti-plasmids with *shi* or *roi* mutations, respectively, resulted in almost normal tumors (Ooms et al. 1981; Binns et al. 1982; Joos et al. 1983 a; Barton et al. 1983). Ooms et al. (1981) hypothesized that the expression of genes 1 and 2 (*Shi*$^+$) results in increased levels of an auxin-like activity, while the expression of gene 4 (*Roi*$^+$) results in increased levels of cytokinin-like activity.

There is evidence that genes 1, 2, and 4 increase the levels of auxin-like and cytokinin-like activities in tumors by directly encoding for enzymes for auxin and cytokinin biosynthesis. Application of α-napthalene acetamide, an intermediate in the biosynthesis of NAA, complemented gene 1 mutation (in the presence of wild-type gene 2), but not a gene 2 mutation (Inze et al. 1984). This result suggests that gene 1 is involved in the synthesis of a precursor, such as α-napthalene acetamide, which is converted by the gene 2 protein to an auxin-like compound, such as NAA. In another approach, gene 2 and gene 4 products were overexpressed in recombinant *E. coli* cells and tested for their ability to catalyze in vitro phytohormone biosynthesis. The product of gene 2 hydrolyzes indole-3-acetamide to the plant hormone auxin, indole-3-acetic acid (IAA) (Schröder et al. 1984). It is believed that the same activity also hydrolyzes α-napthalene acetamide to NAA in tumor cells. The expression of gene 4 product in *E. coli* led to its identification as $\Delta 2$ isopentenyl pyrophosphate: 5′ AMP $\Delta 2$ isopententenyl transferase, which is involved in cytokinin biosynthesis (Barry et al. 1984). The acitivity of the gene 1 product remains to be identified; a significant homology was detected between the predicted primary amino acid sequence of gene 1 and the adenine-binding domain of p-hydroxybenzoate hydroxylase of *Pseudomonas fluorescens* (Klee et al. 1984).

The role of the plant growth hormones, auxin and cytokinin, in tumor formation is also supported by indirect evidence. Skoog and Miller (1957) demonstrated that unorganized tobacco callus can be induced to differentiate into shoots or roots by decreasing or increasing, respectively, the ratio of auxin to cytokinin in the medium. The relative endogenous levels of auxin and cytokinin in tumors broadly correlates with the tumor morphology (Amasino and Miller 1982), including those of tumors incited by Ti-plasmids with *shi* and *roi* mutations (Akiyoshi et al. 1983). Such correlations are not perfect because it is difficult to accurately determine the effective concentration of the phytohormone species that is involved in differentiation, and also because differentiation is dependent on a delicate interplay between several phytohormones. Although there is sufficient suggestive evidence for the role of auxin and cytokinin in tumors, the case is not closed. Not all crown gall tumors have increased levels of auxin and cytokinin (Weiler and Spanier 1981). Ironically, transcripts 1 and 2 which are involved in hormone autotrophy are barely detected in some tumor lines and not at all in other lines (Willmitzer et al. 1983; Joos et al. 1983a). Possibly T-DNA specified hormone biosynthesis is required to initiate tumors but, once initiated tumors could become habituated for growth without hormones (Meins 1982). This possibility is supported by the finding that elevated levels of phytohormones are not necessary for hormone autotrophy of normal habituated tobacco cells (Nakajima et al. 1979).

2.7 Development of Transformed Plant Cells

Crown gall cells, unlike normal plant cells, fail to regenerate intact plants. However, cellular clones of a nopaline teratoma (a tumor with the capacity to organize shoots, but not roots; Turgeon 1982) can be forced, by grafting onto healthy plants, to regenerate normal appearing shoots which can flower and set viable seeds (Braun and Wood 1976; Turgeon et al. 1976; Binns et al. 1981). All vegetative tissues of the normal appearing grafted shoots contain a full complement of T-DNA (Yang et al. 1980; Lemmers et al. 1980), produce nopaline, and when transferred to tissue culture revert to the parental phenotypes of hormone autotrophy and failure to form roots. However, the postmeiotic haploid anther tissue and the selfed-progeny derived from the grafted shoots lack all T-DNA (Yang et al. 1980; Lemmers et al. 1980) and are normal by the same criteria. It is not known if the reversible suppression of some tumor phenotyes in grafted shoots results from an altered expression of T-DNA, from removal of phytohormones by basipetal transport (discussed by Wostemeyer et al. 1984), or from an alternate pathway of morphogenesis.

Teratoma-like tumorous shoots are also formed on tumors incited by octopine Ti-plasmids with *shi* mutations (Ooms et al. 1981; Garfinkel et al. 1981; Leemans et al. 1982; Binns et al. 1982; Hille et al. 1983) and occasionally from protoplasts transformed in vitro with *A. tumefaciens* containing octopine or nopaline Ti-plasmids (Wullems et al. 1981b; Memelink et al. 1983; Fraley and Horsch 1983; Wostemeyer et al. 1984). While most of these shoots are not transformed, some resemble the natural nopaline teratomas not only in their morphology, but also in the presence of the marker opine and in their failure to form roots (Binns et al. 1982; Wullems et al. 1981b; Wöstemeyer et al. 1984). However, these transformed shoots differ from the nopaline teratoma described above in lacking a normal complement of T-DNA, either due to the *shi* mutation (Binns et al. 1982; Hille et al. 1983) or due to spontaneous "deletions" of the left end of T-DNA (Memelink et al. 1983; Wöstemeyer et al. 1984). These "deletions" remove the *Shi* locus, while retaining the *Roi* locus and also retaining the genes for octopine synthase or nopaline synthase. Transformed shoots derived from in vitro protoplast transformation when grafted onto healthy plants produced male-sterile flowers (the male sterility was not due to the presence of T-DNA as even normal plants were male-sterile) (Wullems et al. 1981b; Memelink et al. 1983; Wöstemeyer et al. 1984). Nevertheless, these flowers set viable seeds when cross-fertilized with normal pollen. About half of the resultant progeny seeds gave rise to normal plants and the others to "transformed" seedlings with parental phenotypes, that is, with opines present and unable to form roots. The normal seedlings lacked all T-DNA, whereas the transformed seedlings contained the same T-DNA as the parent. The grafted shoots were heterozygous for the T-DNA, which behaved in a dominant, Mendelian fashion, providing further evidence for the chromosomal location of T-DNA.

It appears that the failure of these teratomas or transformed shoots to regenerate might be related to their inability to form roots. Indeed, Barton et al. (1983) have demonstrated that an insertional mutation in the *Roi* locus of a nopaline Ti-plasmid resulted in attenuated tobacco tumors from which single cell clones can form roots and regenerate into intact, normal appearing, self-fertile plants. The regenerant and its selfed progeny all contained multiple copies of the intact T-DNA with the insertion in the *Roi* locus. In a similar fashion, hairy root cultures can regenerate plants which flower and set seed (David et al. 1984).

Completely normal, flowering plants have also been regenerated from crown gall cells (Turgeon et al. 1976; Sacristan and Melchers 1977; Yang and Simpson 1981; Otten et al. 1981). By placing a cloned nopaline teratoma cell line, called BT37, under conditions that favor regeneration, several independent, normal regenerants have been obtained, which formed roots and set viable seeds. Apparently all had suffered the same large deletion of the central region of the T-DNA during its 5 years in culture, and the BT37 cloned line was a mixture of tumorous and nontumorous cells. The T-DNA fragments in these regenerants were present in all tissues, including those of the selfed progeny plants. Otten et al. (1981) have found that 1 out of about 300 normal shoots arising from tumors incited by an octopine Ti-plasmid with a *shi* mutation, produced octopine and regenerated into a normal flowering plant. All cells of one such regenerant contained octopine as well as a highly truncated T-DNA retaining only the octopine synthase gene (De

Greve et al. 1982). This T-DNA sequence and the octopine-positive phenotype were both transmitted through seeds in a dominant, Mendelian fashion.

These developmental studies show that the presence of certain T-DNA genes in tumor cells inhibits their regeneration as well as their transmission through meiosis and/or germination. Inactivation or deletion of these genes allows regeneration of normal or nearly normal plants and sexual transmission of the T-DNA. As an extension of these studies Ti-plasmids deleted for all of their tumor genes were shown to transform plant cells from which normal flowering plants can be regenerated (Zambryski et al. 1983, 1984; Horsch et al. 1984).

3 Use of Ti-Plasmids as Plant Gene Vectors

3.1 General Remarks

A major motivation for studying the crown gall disease in the past has been its potential as a plant genetic engineering system. Ti-plasmids have two features essential for a transformation vector: the ability to stably replicate (its T-DNA) in transformed cells through its integration into host DNA, and the presence of a selectable marker (hormone-independent growth). Other featurs of Ti-plasmids which are desirable in transformation vectors include: (1) an efficient mechanism for delivering T-DNA into plant cells, (2) a relatively wide host range (De Cleene and De Ley 1976), and (3) no apparent upper limit to the size of foreign sequences that can be introduced into plant cells (Hernalsteens et al. 1980).

The major limitations that have restricted the use of Ti-plasmids as transformation vectors are its tumorigenicity, large size, and host range. As discussed below the first two limitations have been largely overcome. Although Ti-plasmids infect a broad range of dicotyledonous plants, they do not cause tumors in most monocotyledonous plants (De Cleene and De Ley 1976), which include several economically important crops. The basis for this resistance of monocotyledonous plants is unknown. It could be due to several reasons, for example, inability of the bacteria to attach to cell walls or to transfer the T-DNA. Or, if the T-DNA is transferred and expressed, the host may not respond by tumor formation (Binns 1984). Some of these possibilities can now be tested by use of chimeric selectable genes (see below).

3.2 Expression of Foreign Genes in Plants: Construction of Chimeric Selectable Genes

Several foreign genes have been introduced via the Ti-plasmid into plant cells. These include the bacterial genes of transposons Tn5 (Garfinkel et al. 1981) and Tn7 (Hernalsteens et al. 1980), the yeast gene for alcohol dehydrogenase (Barton et al. 1983), the chicken genes for α-actin and ovalbumin (Koncz et al. 1984), and the human genes for β-globin and interferon (unpublished work cited in Caplan et al. 1983). However, expression of these foreign genes in transformed cells was not observed and, where studied, transcripts of these genes were either undetected (Barton et al. 1983; Koncz et al. 1984) or incorrectly initiated (Koncz et al. 1984). However, two foreign plant genes have been correctly transcribed in transformed cells. The gene for the small subunit of ribulose-1,5-bisphosphate carboxylase (RUBPCase) from pea has been introduced into petunia cells where it is properly transcribed, in a light-dependent fashion, is correctly processed, and is translated to yield a functional protein that is transported into chloroplasts (Broglie et al. 1984). Similarly, the gene for phaseolin, a bean storage protein, is properly transcribed and the intervening sequences correctly excised in transformed sunflower tissue (Murai et al. 1983). It appears that plant cells can recognize the transcriptional signals for foreign plant genes, but not for genes from other sources.

In order to express nonplant genes in plant cells, chimeric genes have been constructed in which the nonplant coding sequences are flanked by plant regulatory (noncoding) sequences. The regulatory sequences of the nopaline synthase gene, a T-DNA gene which is expressed in crown gall cells, have been used to express the following bacterial genes: aminoglycoside phosphotransferases, APH(3′)II or APH(3′)I from transposons Tn5 or Tn903, respectively, which inactivate aminoglycosides, such as kanamycin and G-418 (Herrera-Estrella et al. 1983 a; Fraley et al. 1983; Bevan et al. 1983 b), methotrexate-insensitive dihydrofolate reductase in plasmid R67 (Herrera-Estrella et al. 1983 a), the chloramphenicol acetyl transferase, which inactivates chloramphenicol drug (Herrera-Estrella et al. 1983 b), the T-region gene for octopine synthase (Herrera-Estrella et al. 1983 b), and a nopaline synthase: β-galactosidase coding sequence fusion (Helmer et al. 1984). Similarly, the regulatory sequences of octopine synthase gene were used to express an octopine synthase : phaseolin coding sequence fusion (Murai et al. 1983), and the regulatory sequences of the pea small subunit of RuBPCase were used to demonstrate light-inducible expression of the bacterial chloramphenicol acetyl transferase in tobacco cells (unpublished work cited in Caplan et al. 1983). In all cases tested, the foreign gene was properly transcribed and translated into functional proteins (except in the case of phaseolin, which has no known enzyme activity).

While hormone autotrophy of T-DNA has served, and will continue to serve, as a valuable selection marker in studying the integration and expression of both T-DNA and of foreign genes in T-DNA, it is associated with "disease" phenotypes – the inability to regenerate normal plants and the inability of their T-DNA to survive meiosis (see Sect. 2.7). These concerns have evaporated with the demonstration that the Ti-plasmid can be "disarmed" by eliminating the genes involved in hormone autotrophy (Zambryski et al. 1983). This has meant replacing hormone autotrophy with a foreign selectable marker. In the absence of a good plant selectable gene, the chimeric genes mentioned above were tested for their ability to function as dominant selectable markers in transformed cells. Plant cells transformed with chimeric genes in which the regulatory sequences of nopaline synthase gene are fused to the bacterial coding sequences of aminoglycoside phosphotransferases or methotrexate-insensitive dihydrofolate reductase (see above) are able to grow on media containing 50–100 μg ml^{-1} of kanamycin/G-418 or 500 μg ml^{-1} of methotrexate, respectively – levels which kill untransformed plant cells (Herrera-Estrella et al. 1983 a; Fraley et al. 1983; Bevan et al. 1983 b). Horsch et al. (1984) have reported that their vector system can allow the kanamycin resistant transformants to regenerate normal plants which can transmit the foreign gene through seed. The use of disarmed Ti-plasmids (Zambryski et al. 1983, 1984) with such chimeric genes should also allow selection and normal regeneration of transformed plant cells.

3.3 In Vivo Genetic Engineering of the Ti-Plasmid

The problem of manipulating the large (ca. 180 kb) Ti-plasmids in vitro has been circumvented by adapting the same techniques of bacterial genetics which were used successfully in the site-directed mutagenesis of Ti-plasmids (see Sect. 2.6.3). The principle of these techniques (Ruvkun and Ausubel 1981) is simple: DNA manipulations are carried out in *E. coli* using "intermediate" vectors which contain a T-region fragment. The intermediate vectors are then introduced into *A. tumefaciens*, containing a Ti-plasmid, by transformation (Barton and Chilton 1983; deFramond et al. 1983) or by conjugative mobilization with helper plasmids (Comai et al. 1983; Van Haute et al. 1983; Shaw et al. 1983). Under proper selective conditions, or by subsequent conjugation, the intermediate vectors can be forced to cointegrate with the resident Ti-plasmid via the T-region homology. If the engineered DNA is flanked by T-region homologies, a second recombination can lead to homogenotization (Matzke and Chilton 1981; Leemans et al. 1981; Garfinkel et al. 1981; Van Haute et al. 1983; Comai et al. 1983). This strategy has been further simplified by the development of a binary vector system in which the *vir*-region and the T-region are on separate plasmids in *Agrobacterium* cells (deFramond et al. 1983; Hoekema et al. 1983). Presently, the Ti-plasmid can be reduced to a replicon carrying the *vir*-region and a second small shuttle vector containing a plant selectable gene flanked by T-region borders. Genes of interest can be manipulated between the T-region borders in *E. coli,* and the shuttle vector transferred to *A. tumefaciens* containing the *vir*-region replicon, where it is complemented in trans and transferred to plant cells.

3.4 Role of Protoplasts and Tissue Culture in Transformation

Transformation of plant protoplasts with Ti-plasmids has been obtained by several methods outlined below (reviewed by Wullems et al. 1983; Fraley and Horsch 1983; Nagata 1983; Zambryski et al. 1984). The cocultivation method involves incubation of virulent *Agrobacterium tumefaciens* with regenerating protoplasts (Marton et al. 1979; Wullems et al. 1981 a; Fraley et al. 1983). The protoplasts, like the cells in the wound site (see Sect. 2.3), are susceptible at about the same time that they begin to divide. However, it is not known if cocultivation involves the same mechanism of infection as in a wound site. The frequency of hormone autotrophic transformation was between 1–10% of the surviving cells (Wullems et al. 1981 a; Fraley and Horsch 1983). Using kanamycin resistance for selection, the frequency of transformants was reported to be about 6% of the total surviving colonies (Horsch et al. 1984). The in vitro transformants obtained by cocultivation differed from the in planta transformants obtained by stem infection in two respects: (1) in vitro transformants were all hormone autotrophic, whereas the in planta transformants usually are a mixture of transformed and normal cells (Sacristan and Melchers 1977; Binns et al. 1982; Van Slogteren et al. 1983); and (2) in vitro transformants, unlike the in planta transformants, showed a high frequency of segregation of hormone autotrophy and opine production, and a high frequency of shoot-forming transformants (Wullems et al. 1981 b; Fraley and Horsch 1983; Wöstemeyer et al. 1984). In general, however, the T-DNA structures in these two types of transformants were similar (Ooms et al. 1982 b).

Plant protoplasts have also been transformed directly with naked Ti-plasmid DNA in the presence of carrier DNA following incubation with polyethylene glycol and Ca^{2+} (Davey et al. 1980; Krens et al. 1982; Draper et al. 1982). The frequency of transformation was variable and low, about 10^{-4}–10^{-6}. Unlike the in vitro transformants obtained by cocultivation, these transformants appeared to have rearrangements of the T-DNA and lacked the normal ends of T-DNA. Transformation of protoplasts by fusion with *Agrobacterium* spheroplasts (Hasegawa et al. 1981) or by liposome encapsidated Ti-plasmids (Nagata 1983; Fraley and Horsch 1983) were also reported to have a low transformation efficiency of about 10^{-6}.

Infection of stem explants by virulent agrobacteria has been used in several transformation experiments (Barton and Chilton 1983; Bevan et al. 1983 b; Herrera-Estrella 1983 a, b; Helmer et al. 1984). This method is convenient and can be used to obtain stable transformants by selection on hormone-free or drug-containing medium or even by screening for opine production (Zambryski et al. 1983, 1984).

4 Plant Viruses as Potential Transformation Vectors

4.1 General Remarks

Viruses are natural transformation vectors which efficiently introduce, replicate, express, and package their genomes in host cells. In addition, viruses that can spread throughout the plant could potentially be used to systemically transform a whole plant with a foreign gene – especially applicable to host plants lacking a tissue culture system. However, plant viruses are usually associated with pathogenicity and usually have a packaging constraint on the size of foreign genetic material that they can propagate. We do not know which, if any, region of a viral genome can be deleted to eliminate pathogenicity as well as to make room for foreign sequences without affecting its ability to propagate normally. Other limitations of the potential use of most plant viruses include lack of seed transmissibility, restricted host range, difficulty of mechanical innoculation, and insect transmissibility (undesirable because of absence of biological containment of recombinant DNA). Increased knowledge of the molecular biology of plant viruses, the availability of dominant-acting selectable genes for plant transformation (Herrera-Estrella et al. 1983 a; Fraley et al. 1983; Bevan et al. 1983 b), and the development of viral transfections of protoplasts (Maule 1983) should address some of these problems.

Meanwhile, plant transformation vectors could be developed that exploit certain viral features. The most valuable feature of viruses is their ability to replicate autonomously, unlike the T-DNA of Ti-plasmids, which replicates through integration into host chromosomal DNA. The advantage of self-replicating vectors is that they can have a much higher copy number per cell than the integrating vectors and, thus, could provide stronger expression of its selection marker and an increased transformation efficiency. In fact, such autonomous replicons have been the cornerstone of bacterial, yeast, and mammalian transformations (Gunge 1983; Rigby 1983).

The use of different plant viruses as potential plant vectors has been reviewed recently (Howell 1982; Gardner 1983). Since most of the recent progress has been in the molecular biology of cauliflower mosaic virus (CaMV) and since CaMV is the only plant virus that has been used to propagate foreign sequences in plants, it is discussed in detail below. However, other plant viruses, especially the more recently discovered geminiviruses (reviewed by Howarth and Goodman 1982), single-stranded DNA viruses, could prove to be equally useful. Geminiviruses have a bipartite genome and, recently, cloned double-stranded DNA's of the two components of a geminivirus were shown to be infective (Hamilton et al. 1983). Most likely different transformation vectors based on different viruses will be developed, as has happened in the use of animal viruses in mammalian transformation vectors (reviewed by Rigby 1983).

4.2 Cauliflower Mosaic Virus (CaMV)

4.2.1 Structure and Function of the CaMV Genome

Cauliflower mosaic virus (CaMV) is the best characterized member of the Caulimoviruses, a unique group of plant viruses whose genome is double-stranded DNA (see reviews by Howell 1982; Hohn et al. 1982; Hohn and Hohn 1982; Gardner 1983). The icosahedral particles of CaMV contain an open circular DNA, unusual in having ribonucleotides and site-specific single-stranded regions. In most cases, there are three single-stranded regions, two in one strand (the β-strand), and one in the other (the α-strand) (Fig. 3). The single-stranded regions are in fact discontinuities in the DNA strand where one end overlaps the other by 8–43 bp (see Gardner 1983). The function of these features is unknown. While CaMV DNA derived from recombinant clones in *E. coli* lacking these features is infective (see Sect. 4.3), the DNA of viral progeny isolated from such infected plants repossess them.

DNA sequence analysis of three different isolates of CaMV (Franck et al. 1980; Gardner et al. 1981; Balazs et al. 1982) reveals eight tightly packed, open reading frames (ORFs) as potential coding sequences, leaving only two intergenic regions – a short one (ca. 100 bp) between ORFs V and VI and a long one (ca. 1,000 bp) between ORFs VI and VII (Fig. 3). The ORFs are all present on the β-strand, consistent with the observation that only the α-strand is transcribed in

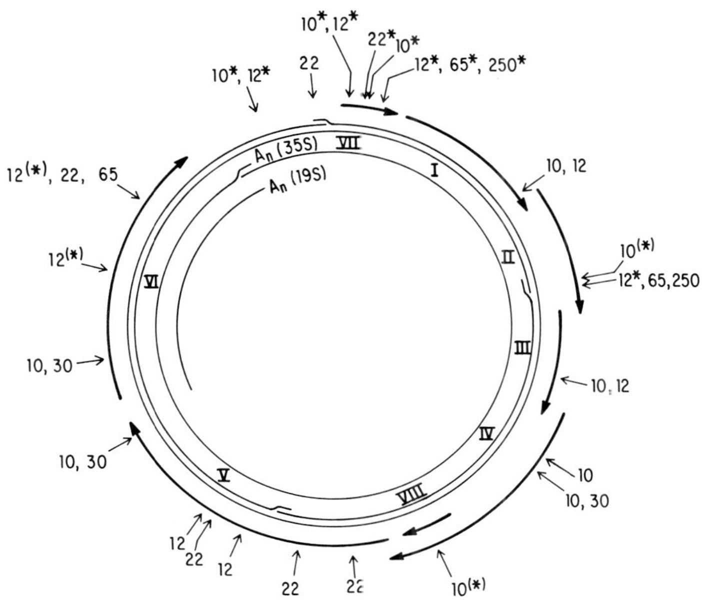

Fig. 3. Transcriptional and genetic maps of the CaMV genome. The three single-stranded regions are shown as short, overlapping strands in the two DNA strands. The *two innermost lines* represent the two major polyadenylated (*An*) transcripts of 19S and 35S. The *open reading frames* (*I* to *VIII*) and their 5′→3′ polarities are shown outside the DNA genome. The location and the size (in bp) of individual insertions are shown outermost. All insertions were lethal to virus except those with the *superscript* *, which were without effect on viral infectivity, and those with a *superscript* (*) which resulted in delayed, but normal symptoms. Data from Daubert et al. (1983), Dixon et al. (1983), and Gardner (1983)

vivo (Guilley et al. 1982). Transcription, which is α-amanitin sensitive, results in two major polyadenylated, capped RNA transcripts (Fig. 3) (Guilfoyle 1980): a 35S species that initiates and terminates in the long, intergenic region and has a 180 bp overlap of its 5′ and 3′ termini, and a 19S subgenomic species that initiates in the short, intergenic region and is 3′ coterminal with the 35S species (Covey and Hull 1981; Guilley et al. 1982; Dudley et al. 1982). Eukaryotic consensus promoter and termination sequences are found upstream of the sites for transcription initiation and termination. Since the ORFs are present in different reading frames, their putative translation products are synthesized individually, apparently from the 35S polycistronic mRNA, and not as a polyprotein precursor. While transcriptionally active covalently closed circular CaMV DNA is located in the nucleus of infected cells, the cellular location of viral replication is unknown. It has been proposed that CaMV replicates via reverse transcription (Pfeiffer and Holn 1983; Guilley et al. 1983; Hull and Covey 1983).

4.2.2 Mutagenesis Studies

Mutagenesis of the CaMV genome by random and site-specific insertions/deletions has been used to identify the genetic determinants of viral infectivity, to identify dispensable regions for inserting foreign DNA, and to determine the size limit of the insertions (Howell et al. 1981; Gronenborn et al. 1982; Lebeurier et al. 1980, 1982; Dixon et al. 1983; Daubert et al. 1983). The results of these studies, some of which are summarized in Fig. 3, show that short insertions or deletions in all ORFs, except ORFs II and VII, are lethal to the virus or cause delayed viral symptoms on infected plants. Therefore, assuming no polar effect of the insertion mutations, the ORFs I, III, IV, V, and VI are essential for infection. There is evidence that ORF IV encodes the viral capsid protein (Daubert et al. 1982) and that ORF VI a protein of the viroplasm – the cytoplasmic matrix in which the virions are embedded in infected cells (Odell and Howell 1980; Xiong et al. 1982). No gene product has been assigned to ORFs I, III, and V, although ORF V, which is the most conserved region between different viral isolates, has a predicted protein sequence which has homology to animal virus reverse transcriptase (Toh et al. 1983).

The ORFs II and VII and the long, intergenic region can tolerate short insertions. Insertions in ORF II, which encodes an aphid-transmissibility factor (Armour et al. 1983; Wollston et al. 1983; Daubert et al. 1983; Gardner 1983; Givord et al. 1984) and in ORF VII, with no known function, result in different phenotypes depending on the insertion and the viral strain used. In an aphid-transmissible strain an in-frame 12 bp insertion in ORF II resulted in normal infection, whereas frameshift insertions in the same site of 10 bp, 65 bp, or 250 bp resulted in delayed symptoms or loss of infectivity (Dixon et al. 1983; Daubert et al. 1983). On the other hand, a nonaphid-transmissible mutant of CaMV could tolerate insertions of 65 bp or 256 bp, but not of 531 or 1200 bp in the same site of ORF II (Gronenborn et al. 1981). Similarly, in some strains insertions of up to 270 bp (larger-sized inserts were not tested) in ORF VII did not affect their viability, while in a strain with a natural deletion of 421 bp insertions of 120–550 bp, but not a deletion of 105 bp, in ORF VII destroyed infectivity (Howell et al. 1981; Gronenborn et al. 1981; Gardner 1983). These results suggest that insertions could have pleiotropic effects in addition to their contribution to the size of the viral genome. For example, a frameshift mutation could introduce a long, intergenic region which could affect the translation of the downstream sequence on a polycistronic mRNA (Dixon et al. 1983).

Another region that can tolerate at least short insertions is the long, intergenic region – only one of several insertions in this region was lethal to the virus. This insertion was made 45 bp upstream of the single-stranded site (Fig. 3) and possibly affected viral replication (Dixon et al. 1983).

The apparent severity of size limitation of inserts in the CaMV genome may be related to its packaging in the virus particles. In order to overcome this limitation attempts have been made to develop a "helper virus" system in which certain viral genes are removed to accomodate foreign sequences and provided in trans (Howell et al. 1981; Lebeurier et al. 1982; Daubert et al. 1983). Coinfection of host plants with pairs of mutant viral DNA that were individually noninfective resulted in normal infection. However, this rescue of defective genomes occurred not by complementation, but by recombination between the viral genomes (Walden and Howell 1982). Gardner (1983) proposed that systemic infection necessary for the viral symptoms might select for recombination over complementation.

4.2.3 Prospects for Using CaMV as a Transformation Vector

The attractions of CaMV as a potential transformation vector include (1) its short (8 kb), completely sequenced, circular DNA, (2) its rapid and systemic infection to a high titer, and (3) the infectivity of the viral DNA. It has also been demonstrated that a recombinant clone of CaMV made in *E. coli*, is infective following release of the vector DNA (Howell et al. 1980; Hohn et al. 1982) or without prior release of the vector DNA, if the viral DNA is flanked by homologous viral DNA arms (Lebeurier et al. 1982; Walden and Howell 1983). Although short, foreign sequences (up to 270 bp) inserted in nonessential regions of the CaMV genome have been propagated in infected plants, the biggest obstacle in the development of CaMV as a vector is the apparently strict limitation to the size of "passenger" DNA that it can tolerate. The upper size limit for viral packaging must be accurately assessed and this limitation must be overcome by developing a helper virus system in which homologous recombination is not possible. Other limitations of a CaMV as a vector include its narrow host range – largely limited to members of the family Cruciferae, and its nontransmissibility through seeds (see review by Gardner 1983).

It is important to emphasize that the assay for CaMV infectivity used in mutagenesis studies depends on proper viral replication, expression, packaging, and cell-to-cell spread. Some mutations, especially the large insertions in nonessential regions, may allow the replication of the mutant virus without showing any viral symptom. Therefore, it may be possible to exploit the replication mechanism of CaMV to propagate larger foreign sequences in protoplasts from susceptible hosts. While protoplasts of several *Brassica* species have been transfected with CaMV virus (Howell and Hull 1978; Furosawa et al. 1980; Maule 1983) and CaMV DNA (Yamaoka et al. 1982), there have been no reports of protoplast transfection with cloned CaMV DNA. The availability of dominant selectable genes for plant cells (see Sect. 3.2) should help in developing a protoplast system for CaMV replication. Such a system could be analogous to other autonomously replicating genetic elements, such as plasmids in bacteria, the 2 μ circle and the autonomously replicating sequences (ars) in yeast (Gunge 1983), and the SV40 viral system in mammalian cells (Rigby 1983).

5 Concluding Remarks

The recent progress in our understanding of the molecular basis of Ti-plasmid induced crown gall tumorigenesis has made it possible to use the Ti-plasmid to genetically engineer plants with foreign genes that are stably expressed and inherited in a Mendelian fashion. This, in conjunction with the capacity of some plant cells to regenerate into whole plants, provides us with an opportunity to study the nucleotide sequences involved in the control of expression of introduced plant genes during differentiation. Much more must be learned about the expression of foreign genes before we can genetically engineer desirable traits into crop plants. We already know from the study of Ti-plasmid genes in tumors that foreign gene expression can vary greatly. The availability of dominant selectable genes for plants and appropriate protoplast transformation systems should allow improvements in the Ti-plasmid based vectors as well as development of novel vectors based on the replication mechanism of plant viruses, such as CaMV.

Acknowledgments. I wish to thank my colleagues Deborah Chaleff, Ron Hoess, Kenneth Leto, and Barbara Mazur for critically reading the manuscript, Andrew Binns for discussions and Pat Pulcher for excellent assistance in preparing the manuscript. This chapter reviews publications prior to June 1984.

References

Akiyoski DE, Morris RO, Hinz R, Mischke BS, Kosuge T, Garfinkel DJ, Gordon MP, Nester EW (1983) Cytokinin/Auxin balance in crown gall tumors is regulated by specific loci in the T-DNA. Proc Natl Acad Sci USA 80:407–411

Albinger G, Beiderbeck R (1977) Übertragung der Fähigkeit zur Wurzelinduktion von *Agrobacterium rhizogenes* auf *A. tumefaciens*. Phytopathol Z 90:306–310

Amasino RM, Miller CO (1982) Hormonal control of tobacco crown gall tumor morphology. Plant Physiol (Bethesda) 69:389–392

Armour SL, Melcher U, Pirone TP, Lyttle DJ, Essenberg RC (1983) Helper component for aphid transmission encoded by region II of cauliflower mosaic virus DNA. Virology 129:25–30

Balazs E, Guilley H, Jonard G, Richards K (1982) Nucleotide sequence of DNA from an altered-virulence isolate D/H of cauliflower mosaic virus. Gene (Amst) 19:239–249

Banerjee D, Basu M, Choudhary I, Chatterjee GC (1981) Cell surface carbohydrates of *Agrobacterium tumefaciens* involved in adherence during crown gall tumor initiation. Biochem Biophys Res Commun 100:1384–1388

Barker RF, Idler KB, Thompson DV, Kemp JD (1983) Nucleotide sequence of the T-DNA region from the *Agrobacterium tumefaciens* octopine Ti plasmid pTil5955. Plant Mol Biol 2:335–350

Barry GF, Roger SG, Fraley RT, Brand L (1984, in press) Identification of a cloned cytokinin biosynthetic gene. Gene (Amst)

Barton KA, Chilton MD (1983) *Agrobacterium* Ti-plasmids as vectors for plant genetic engineering. In: Wu R, Grossman L, Moldave K (eds) Recombinant DNA, part C. Methods in Enzymology, Vol 101. Academic, New York, pp 527–539

Barton KA, Binns AN, Matzke AJM, Chilton M-D (1983) Regeneration of intact tobacco plants containing full length copies of genetically engineered T-DNA, and transmission of T-DNA to R_1 progeny. Cell 32:1033–1043

Benoist C, O'Hare K, Breathnach R, Chambon P (1980) The ovalbumin gene sequence of putative control regions. Nucleic Acids Res 8:127–142

Bevan M, Chilton M-D (1982a) Multiple transcripts of T-DNA detected in nopaline crown gall tumors. J Mol Appl Genet 1:539–546

Bevan MW, Chilton M-D (1982b) T-DNA of the *Agrobacterium* Ti and Ri plasmids. Annu Rev Genet 16:357–384

Bevan M, Barnes WM, Chilton M-D (1983a) Structure and transcription of the nopaline synthase gene region of T-DNA. Nucleic Acids Res 11:369–385

Bevan MW, Flavell RB, Chilton M-D (1983b) A chimeric antibiotic resistance gene as a selectable marker for plant cell transformation. Nature (Lond) 304:184–184

Binns AN (1984, in press) The biology and molecular biology of plant cells infected by *Agrobacterium tumefaciens*. In: Miffin B (ed) Oxford surveys of plant molecular and cell biology, vol 1. Oxford Univ Press, London

Binns AN, Wood HN, Braun AC (1981) Suppression of the tumorous state in crown gall teratomas of tobacco: A clonal analysis. Differentiation 19:97–102

Binns AN, Sciaky D, Wood HN (1982) Variation in hormone autonomy and regenerative potential of cells transformed by strain A66 of *Agrobacterium tumefaciens*. Cell 31:605–612

Braun AC (1952) Conditioning of the host cell as a factor in the transformation process. Growth 16:65–74

Braun AC (1982) A history of the crown gall problems. In: Kahl G, Schell J (eds) Molecular biology of plant tumors. Academic, New York, pp 155–210

Braun AC, Wood HN (1976) Suppression of the neoplastic state with the acquisition of specialized functions in cells, tissues, and organs of crown gall teratomas of tobacco. Proc Natl Acad Sci USA 73:496–500

Breathnach R, Chambon P (1981) Organization and expression of eukaryotic split genes coding for proteins. Annu Rev Biochem 50:349–383

Broglie R, Coruzzi G, Fraley RT, Rogers SG, Horsch RB, Niedermeyer JG, Fink CL, Flick JS, Chua N-H (1984) Light-regulated expression of a pea ribulose-1,5-bisphosphate carboxylase small subunit gene in transformed plant cells. Science (Wash DC) 224:838–843

Calos MP, Miller JH (1980) Transposable elements. Cell 20:579–595

Caplan A, Herrera-Estrella L, Inze D, Van Haute E, Van Montagu M, Schell J, Zambryski P (1983) Introduction of genetic material into plant cells. Science (Wash DC) 222:815–821

Chilton M-D, Drummond MH, Merlo DJ, Sciaky D, Montoya AL, Gordon MP, Nester EW (1977) Stable incorporation of plasmid DNA into higher plant cells: The molecular basis of crown gall tumorigenesis. Cell 11:263–271

Chilton M-D, Drummond MH, Merlo DJ, Sciaky D (1978) Highly conserved DNA of Ti plasmids overlaps T-DNA maintained in plant tumors. Nature (Lond) 275:147–149

Chilton M-D, Tepfer DA, Petit A, David CC, Delbert F, Tempe J (1982) *Agrobacterium rhizogenes* inserts T-DNA into plant roots. Nature (Lond) 295:432–434

Chilton WS, Tempe J, Matzke M, Chilton M-D (1984) Succinamopine: a new crown gall opine. J Bacteriol 157:357–362

Comai L, Schilling-Cordaro C, Mergia A, Houck CH (1983) A new technique for genetic engineering of *Agrobacterium* Ti-plasmid. Plasmid 10:21–30

Covey SN, Hull R (1981) Transcription of cauliflower mosaic virus DNA: Detection of transcripts, properties and location of the gene encoding the virus inclusion body protein. Virology 11:463–474

Daubert S, Richins R, Shepherd RJ, Gardner RC (1982) Mapping of the coat protein gene of cauliflower mosaic virus by its expression in a prokaryotic system. Virology 122:444–449

Daubert S, Shepherd RJ, Gardner RC (1983) Insertional mutagenesis of the cauliflower mosaic virus genome. Gene (Amst) 25:201–208

Davey MR, Cocking EC, Freeman N, Pearce N, Tudor I (1980) Transformation of *Petunia* protoplasts by isolated *Agrobacterium* plasmid. Plant Sci Lett 18:307

David C, Chilton M-D, Tempe J (1984) Conservation of T-DNA in plants regenerated from hairy root cultures. Biotechnology 2:73–76

De Beuckleer M, Lemmers M, De Vos G, Willmitzer L, Van Monntagu M, Schell J (1981) Further insight on the transferred-DNA of octopine crown gall. Mol Gen Genet 183:283–288

De Cleene M, De Ley J (1976) The host range of crown gall. Bot Rev 42:389–466

De Cleene M, De Ley JD (1981) The host range of infectious hairy-root. Bot Rev 47:147–194

de Framond A, Barton K, Chilton MD (1983) Mini-Ti: A new strategy for plant genetic engineering. Biotechnology 1:262–269

De Greve H, Decraemer H, Seurinck J, Van Montagu M, Schell J (1981) The functional organization of the octopine *Agrobacterium tumefaciens* plasmid pTiB6S3. Plasmid 6:235–248

De Greve H, Leemans J, Hernalsteens JP, Thia-Toong L, De Beuckeleer L, Willmitzer L, Otten L, Van Moutagu M, Schell J (1982) Regeneration of normal and fertile plants that express octopine synthasa, from tobacco crown galls after deletion of tumor-controlling functions. Nature (Lond) 300:752–755

De Greve H, Dhaese P, Seurinck J, Lemmers M, Van Montagu M, Schell J (1983) Nucleotide sequence and transcript map of the *Agrobacterium tumefaciens* Ti-plasmid-encoded octopine synthase gene. J Mol Appl Genet 1:499–511

Depicker A, Stachel S, Dhaese P, Zambryski P, Goodman HM (1982) Nopaline synthase: transcript mapping and DNA sequence. J Mol Appl Genet 1:561–574

Depicker A, Van Montagu M, Schell J (1983) Plant cell transformation by *Agrobacterium* plasmids. In: Kosuge T, Meredith CP, Hollaender A (eds) Genetic engineering of plants. Plenum, New York, pp 143–176

Dhaese P, De Greve H, Gielen J, Seurinck J, Van Montagu M, Schell J (1983) Identification of sequences involved in the polyadenylation of higher plant nuclear transcripts using *Agrobacterium* T-DNA genes as models. EMBO J 2:419–426
Dixon LK, Koenig I, Hohn T (1983) Mutagenesis of cauliflower mosaic virus. Gene (Amst) 25:189–199
Draper J, Davey MR, Freeman JP, Cocking EC, Cox BJ (1982) Ti-plasmid homologous sequences present in tissues from *Agrobacterium* plasmid-transformed *Petunia* protoplasts. Plant Cell Physiol 23:451–458
Dudley KR, Odell JT, Howell SH (1982) Structure and 5′-termini of the large and 19S RNA transcripts encoded by the cauliflower mosaic virus genome. Virology 117:19–28
Ellis JG, Murphy PH (1981) Four new opines from crown gall tumors – their detection and properties. Mol Gen Genet 181:36–43
Engler G, Depicker A, Maenhout R, Villarroel R, Van Montagu M, Schell J (1981) Physical mapping of DNA base sequence homologies between an octopine and a nopaline Ti-plasmid of *Agrobacterium tumefaciens*. J Mol Biol 152:183–208
Fraley RT, Horsch RB (1983) In vitro plant transformation systems using liposomes and bacterial co-cultivation. In: Kosuge T, Meredith CP, Hollaender A (eds) Genetic engineering of plants. Plenum, New York, pp 177–194
Fraley RT, Rogers SG, Horsch RB, Sanders PR, Flick JS, Adams SP, Bittner ML, Brand LA, Fink CL, Fry JS, Gallupi GR, Goldberg SB, Hoffmann NL, Woo SC (1983) Expression of bacterial genes in plant cells. Proc Natl Acad Sci USA 80:4803–4807
Franck A, Guilley H, Jonard G, Richards K, Hirth L (1980) Nucleotide sequences of cauliflower mosaic virus DNA. Cell 21:285–294
Furusawa I, Yamaoka N, Okuno T, Yamamoto M, Kohno M, Kunoh H (1980) Infection of turnip protoplasts with cauliflower mosaic virus. J Gen Virol 48:431–435
Gardner RC (1983) Plant viral vectors: CaMV as an experimental tool. In: Kosuge T, Meredith CP, Hollaender A (eds) Genetic engineering of plants. Plenum, New York, pp 121–142
Gardner R, Howarth A, Hahn P, Brown-Leudi M, Shepherd R, Messing J (1981) The complete nucleotide sequence of an infectious clone of cauliflower mosaic virus by M13mp7 shotgun sequencing. Nucleic Acids Res 9:2871–2888
Garfinkel DJ, Nester EW (1980) *Agrobacterium tumefaciens* mutants affected in crown gall tumorigenesis and octopine catabolism. J Bacteriol 144:732–743
Garfinkel DJ, Simpson RB, Ream LW, White FF, Gordon MP, Nester EW (1981) Genetic analysis of crown gall: fine structure map of the T-DNA by site directed mutagenesis. Cell 27:143–153
Gelvin SB, Thomashow MF, McPherson JC, Gordon MP, Nester EW (1982) Sizes and map positions of several plasmid DNA-encoded transcripts in octopine-type crown gall tumors. Proc Natl Acad Sci USA 79:76–80
Gelvin SB, Karchner SJ, DiRita VJ, Talierco EW (1983) Transcription of the Ti-plasmid in crown gall tumors. In: Pühler A (ed) Molecular genetics of the bacteria-plant interaction. Springer, Berlin Heidelberg New York, pp 292–302
Gielen J, De Beuckleer M, Seurinck J, Deboeck F, DeGreve H, Lemmers M, Van Montagu M, Schell J (1984) The complete nucleotide sequence of the TL-DNA of the *Agrobacterium tumefaciens* plasmid pTiAch5. EMBO J 3:835–846
Givord L, Xiong C, Giband M, Koenig I, Hohn T, Lebeurier G, Hirth L (1984) A second cauliflower mosaic virus gene product influences the structure of the viral inclusion body. EMBO J 3:1423–1427
Gronenborn B, Gardner R, Schaefer S, Shepherd R (1981) Propagation of foreign DNA in plants using CaMV as a vector. Nature (Lond) 294:773–775
Guilfoyle TJ (1980) Transcription of the cauliflower mosaic virus genome in isolated nuclei from turnip leaves. Virology 107:71–80
Guilley H, Dudley RK, Jonard G, Balazs E, Richards KE (1982) Transcription of cauliflower mosaic virus DNA: detection of promoter sequences, and characterization of transcripts. Cell 30:763–773
Guilley H, Richards KE, Jonard G (1983) Observations concerning the discontinuous DNA's of cauliflower mosaic virus. EMBO J 2:277–282
Gunge N (1983) Yeast DNA plasmids. Annu Rev Microbiol 37:253–276

Hamilton WDO, Bisaro DM, Coutts RHA, Buck KW (1983) Demonstration of the bipartite nature of the genome of a single-stranded DNA plant virus by infection with the cloned DNA components. Nucleic Acids Res 11:7387–7396

Hasezawa S, Nagata T, Syono K (1981) Transformation of *Vinca* protoplasts midiated by *Agrobacterium* spheroplasts. Mol Gen Genet 182:206–210

Helmer G, Casadaban M, Bevan M, Kayes L, Chilton M-D (1984) A new chimeric gene as a marker for plant transformation: the expression of *Escherichia coli* β-galactosidase in sunflower and tobacco cells. Biotechnology 2:500–507

Hepburn AG, Clarke LE, Blundy KS, White J (1983a) Nopaline Ti-plasmid, pTiT37, T-DNA insertions into a flax genome. J Mol Appl Genet 2:211–224

Hepburn AG, Clarke LE, Pearson L, Blundy KS, White J (1983b) The fate of T-DNA in flax. In: Chater KF, Cullis CA, Hopwood DA, Johnston AAWB, Woolhouse HW (eds) Genetic rearrangement. Sinauer, Sunderland, Mass, pp 169–181

Hernalsteens JP, Van Vliet F, De Beuckleer M, Depicker A, Engler G, Lemmers M, Holsters M, Van Montagu M, Schell J (1980) The *Agrobacterium tumefaciens* Ti-plasmid as a host vector system for introducing foreign DNA in plant cells. Nature (Lond) 287:654–656

Herrera-Estrella L, De Block M, Messens E, Hernalsteens J-P, Van Montagu M, Schell J (1983a) Chimeric genes as dominant selectable markers in plant cells. EMBO J 2:987–995

Herrera-Estrella L, Depicker A, Van Montagu M, Schell J (1983b) Expression of chimeric genes transferred into plant cells using a Ti-plasmid-derived vector. Nature (Lond) 303:209–213

Hille J, Klasen I, Schilperoort R (1982) Construction and application of R prime plasmids, carrying different segments of an octopine Ti-plasmid from *Agrobacterium tumefaciens* for complementation of *vir* genes. Plasmid 7:107–118

Hille J, Wullems G, Schilperoort R (1983) Non-oncogenic T-region mutants of *Agrobacterium tumefaciens* do transfer T-DNA into plant cells. Plant Mol Biol 2:155–163

Hille J, Van Kan J, Schilperoort R (1984) *Trans*-acting virulence functions of the octopine Ti plasmid from *Agrobacterium tumefaciens*. J Bacteriol 158:754–756

Hoekema A, Hirsch PR, Hooykaas PJJ, Schilperoort RA (1983) A binary plant vector strategy based on separation of *vir*-and T-region of the *Agrobacterium tumefaciens* Ti-plasmid. Nature (Lond) 303:179–180

Hoekema A, Hooykaas PJ, Schilperoort RA (1984) Transfer of the octopine T-DNA segment to plant cells mediated by different types of *Agrobacterium* tumor- or root-inducing plasmids: generality of virulence systems. J Bacteriol 158:383–385

Hohn B, Hohn T (1982) Cauliflower mosaic virus: a potential vector for plant genetic engineering. In: Kahl G, Schell J (eds) Molecular biology of plant tumors. Academic, New York, pp 549–560

Hohn T, Richards K, Lebeurier G (1982) Cauliflower mosaic virus on its way to becoming a useful plant vector. In: Hofschneider PH, Goebel W (eds) Gene cloning in organisms other than *E. coli*. Curr Top Microbiol Immunol, vol 96. Springer, Berlin Heidelberg New York, pp 193–236

Holsters M, Silva B, Van Vliet F, Genetello C, De Block M, Dhaese P, Depicker A, Inze D, Engler G, Villarroel R, Van Montagu M, Schell J (1980) The functional organization of the nopaline *A. tumefaciens* plasmid pTiC58. Plasmid 3:212–230

Holsters M, Villaroel R, Van Montagu M, Schell J (1982) The use of selectable markers for the isolation of plant DNA/T-DNA junction fragments in a vector. Mol Gen Genet 185:283–289

Holsters M, Villarroel R, Gielen J, Seurinck J, De Greve H, Van Montagu M, Schell J (1983) An analysis of the boundaries of the octopine TL-DNA in tumors induced by *Agrobacterium tumefaciens*. Mol Gen Genet 190:35–41

Hooykaas PJJ, Schilperoort RA (1984) The molecular genetics of crown gall tumorigenesis. In: Scandalios JG (ed) Molecular genetics of plants. Adv Genet, vol 22. Academic Press, Orlando, pp 209–283

Horsch RB, Fraley RT, Rogers SG, Sanders PR, Lloyd A, Hoffmann N (1984) Inheritance of functional foreign genes in plants. Science (Wash DC) 223:496–498

Howarth AJ, Goodman RM (1982) Plant viruses with genomes of single-stranded DNA. Trends Biochem Sci 7:180–182

Howell SH (1982) Plant molecular vehicles: potential vectors for introducing foreign DNA into plants. Annu Rev Plant Physiol 33:609–650

Howell SH, Hull R (1978) Replication of cauliflower mosaic virus and transcription of its genome in turnip leaf protoplasts. Virology 86:468–481

Howell SH, Walker LL, Dudley RK (1980) Cloned cauliflower mosaic virus DNA infects turnips (*Brassica rapa*). Science (Wash DC) 208:1265–1267

Howell S, Walker L, Walden R (1981) Rescue of in vitro generated mutants of cloned CaMV genome in infected plants. Nature (Lond) 293:485–486

Hull R, Covey SN (1983) Characterization of cauliflower mosaic virus DNA forms isolated from infected turnip leaves. Nucleic Acids Res 11:1881–1895

Inze D, Follin A, Van Lijsebettens M, Simeons C, Gentello C, Van Montagu M, Schell J (1984) Genetic analysis of the individual T-DNA genes of *Agrobacterium tumefaciens;* further evidence that two genes are involved in indole-3-acetic acid synthesis. Mol Gen Genet 194:265–274

Iyer UN, Klee HJ, Nester EW (1982) Units of genetic expression in the virulence region of a plant tumor-inducing plasmid of *Agrobacterium tumefaciens.* Mol Gen Genet 188:418–424

Joos H, Inze D, Caplan A, Sormann M, Van Montagu M, Schell J (1983a) Genetic analysis of T-DNA transcripts in nopaline crown galls. Cell 32:1057–1067

Joos H, Timmerman B, Van Montagu M, Schell J (1983b) Genetic analysis of transfer and stabilization of *Agrobacterium* DNA in plant cells. EMBO J 2:2151–2160

Kahl G, Schell J (eds) (1982) Molecular biology of plant tumors. Academic, New York

Karchner SJ, DiRita VJ, Gelvin SB (1984) Transcript analysis of T_RDNA in octopine-type crown gall tumors. Mol Gen Genet 194:159–165

Kemp JD (1982) Enzymes in octopine and nopaline metabolism. In: Kahl G, Schell J (eds) Molecular biology of plant tumors. Academic, New York, pp 461–474

Kemp JD (1983) Genetic engineering of plants by novel approaches. In: Inouye M (ed) Experimental manipulation of gene expression. Academic, New York, pp 119–135

Kerr A, Ellis JG (1982) Conjugation and transfer of Ti-plasmids in *Agrobacterium tumefaciens.* In: Kahl G, Schell J (eds) Molecular biology of plant tumors. Academic, New York, pp 321–344

Klapwijk PM, Scheulderman T, Schilperoort RA (1978) Coordinate regulations of octopine degradation and conjugative transfer of Ti-plasmids in *Agrobacterium tumefaciens:* evidence for a common regulatory gene and separate operons. J Bacteriol 136:775–785

Klee HJ, Gordon MP, Nester EW (1982) Complementation analysis of oncogenicity. J Bacteriol 150:327–331

Klee H, Montoya A, Horodyski F, Lichtenstein C, Garfinkel D, Fuller S, Flores C, Peschon J, Nester E, Gordon M (1984) Nucleotide sequence of the *tms* genes of the pTiA6NC octopine Ti-plasmid: two gene products involved in plant tumorigenesis. Proc Natl Acad Sci USA 81:1728–1732

Koekman BP, Ooms G, Klapwijk PM, Schilperoort RA (1979) Genetic map of an octopine Ti-plasmid. Plasmid 2:347–357

Koekman BP, Hooykaas PJJ, Schilperoort RA (1982) A functional map of the replicator region of the octopine Ti-plasmid. Plasmid 7:119–132

Koncz C, De Greve H, Andre D, Deboeck F, Van Montagu M, Schell J (1983) The opine synthase genes carried by Ti-plasmids contain all signals necessary for expression in plants. EMBO J 2:1597–1603

Koncz C, Kreuzaler F, Kalman ZS, Schell J (1984) A simple method to transfer, integrate and study expression of foreign genes, such as chicken ovalbumin and α-actin in plant tumors. EMBO J 5:1029–1037

Krens FA, Molendijk L, Wullens GJ, Schilperoort RA (1982) In vitro transformation of plant protoplasts with Ti-plasmid DNA. Nature (Lond) 296:72–74

Lahners K, Byrne MC, Chilton M-D (1984) T-DNA fragments of hairy root plasmid pRi8196 are distantly related to octopine and nopaline Ti-plasmid T-DNA. Plasmid 11:130–140

Landy A, Ross W (1977) Viral integration and excision: structure of the lambda att sites. Science (Wash DC) 197:1147–1159

Lebeurier G, Hirth L, Hohn T, Hohn B (1980) Infectivities of native and cloned DNA of cauliflower mosaic virus. Gene (Amst) 12:139–146

Lebeurier G, Hirth L, Hohn B, Hohn T (1982) In vivo recombination of cauliflower mosaic virus DNA. Proc Natl Acad Sci USA 79:2932–2936

Leemans J, Shaw C, Deblaere R, De Greve H, Hernalsteens JP, Maes M, Van Montagu M, Schell J (1981) Site-specific mutagenesis of *Agrobacterium* Ti-plasmids and transfer of genes to plant cells. J Mol Appl Genet 1:149–164

Leemans J, Deblaere R, Willmitzer L, De Greve H, Hernalsteens JP, Van Montagu M, Schell J (1982) Genetic identification of functions of TL-DNA transcripts in octopine crown galls. EMBO J 1:147–152

Lemmers M, De Beuckleer M, Holsters M, Zambryski P, Depicker A, Hernalsteens JP, Van Montagu M, Schell J (1980) Internal organization, boundaries and integration of Ti-plasmid DNA in nopaline crown gall tumors. J Mol Biol 144:353–376

Lichtenstein C, Klee H, Montoya A, Garfinkel D, Fuller S, Flores C, Nester E, Gordon M (1984) Nucleotide sequence and transcript mapping of the *tmr* gene of the pTiA6NC octopine Ti-plasmid: a bacterial gene involved in plant tumorigenesis. J Mol Appl Genet 2:354–362

Lipetz J (1966) Crown gall tumorigenesis. II. Relations between wound healing and the tumorigenic response. Cancer Res 26:1597–1604

Liu ST, Perry KL, Schandl CL, Kado CI (1982) *Agrobacterium* Ti-plasmid indoleacetic acid gene is required for crown gall oncogenesis. Proc Natl Acad Sci USA 79:2812–2816

Lundquist RC, Close TJ, Kado CL (1984) Genetic complementation of *Agrobacterium tumefaciens* Ti plasmid mutants in the virulence region. Mol Gen Genet 193:1–7

Marton L, Wullems GJ, Molendijk L, Schilperoort RA (1979) In vitro transformation of cultured cells from *Nicotiana tabacum* by *Agrobacterium tumefaciens*. Nature (Lond) 277:129–131

Matthysee AG (1983) The use of tissue cultures in the study of crown gall and other bacterial diseases. In: Hegelson JP, Deverall BJ (eds) Use of tissue culture and protoplasts in plant pathology. Academic, Sydney, pp 51–68

Matzke AJM, Chilton M-D (1981) Site-specific insertion of genes into T-DNA of the *Agrobacterium* tumor-inducing plasmid: an approach to genetic engineering of higher plant cells. J Mol Appl Genet 1:39–49

Maule AJ (1983) Infection of protoplasts from several *Brassica* species with cauliflower mosaic virus following inoculation using polyethylene glycol. J Gen Virol 64:2555–2660

McPherson JC, Nester EW, Gordon MP (1980) Proteins encoded by *Agrobacterium tumefaciens* Ti-plasmid DNA (T-DNA) in crown gall tumors. Proc Natl Acad Sci USA 77:2666–2670

Meins F (1982) Habituation of cultured plant cells. In: Kahl G, Schell J (eds) Molecular biology of plant tumors. Academic, New York, pp 3–31

Memelink J, Wullems GJ, Schilperoort RA (1983) Nopaline T-DNA retained during regeneration and generative propagation of transformed plants. Mol Gen Genet 190:516–522

Murai N, Kemp JD (1982a) T-DNA of pTil5955 from *Agrobacterium tumefaciens* is transcribed into a minimum of seven polyadenylated RNAs in a sunflower crown gall tumor. Nucleic Acids Res 10:1679–1689

Murai N, Kemp JD (1982b) Octopine synthase messenger RNA isolated from sunflower crown gall callus is homologous to the Ti-plasmid of *Agrobacterium tumefaciens*. Proc Natl Acad Sci USA 79:86–90

Murai N, Sutton D, Murray M, Slightom J, Merlo D, Reichert N, Sengupta-Gopalan C, Stock C, Barker R, Kemp J, Hall T (1983) Phaseolin gene from bean is expressed after transfer to sunflower via tumor-inducing plasmid vectors. Science (Wash DC) 222:476–481

Nagata T (1983) Liposomes as a carrier of Ti-plasmid into protoplasts. In: Pühler A (ed) Molecular genetics of the bacteria-plant interaction. Springer, Berlin Heidelberg New York, pp 268–273

Nakajima H, Yakota T, Matsumoto T, Noguchi M, Takahashi N (1979) Relationship between hormone content and autonomy in various autonomous tobacco cells cultured in suspension. Plant Cell Physiol 29:1489–1499

Odell J, Howell S (1980) Identification, mapping and characterization of mRNA for P66, a CaMV-coded protein. Virology 102:439–359

Ooms G, Klapwijk PM, Poulis JA, Schilperoort RA (1980) Characterization of Tn904 insertions in octopine Ti-plasmid mutants of *Agrobacterium tumefaciens*. J Bacteriol 144:82–91

Ooms G, Hooykaas PJJ, Moolenaar G, Schilperoort RA (1981) Crown gall plant tumors of abnormal morphology, induced by *Agrobacterium tumefaciens* carrying mutated octopine Ti plasmids; analysis of T-DNA functions. Gene (Amst) 14:33–50

Ooms G, Bakker A, Molendijk L, Wullems GJ, Gordon MP, Nester EW, Schilperoort RA (1982a) T-DNA organization in homogeneous and heterogeneous octopine-type crown gall tissues of *Nicotiana tabacum*. Cell 30:589–597
Ooms G, Mulendijk L, Schilperoort RA (1982b) Double infection of tobacco plants two complementing octopine T-region mutants of *Agrobacterium tumefaciens*. Plant Mol Biol 1:217–226
Otten LABM, Vreugdenhil D, Schilperoort RA (1977) Properties of D(+) lysopine dehydrogenase from crown gall tumour tissue. Biochim Biophys Acta 485:268–277
Otten L, De Greve H, Hernalsteens JP, Van Montagu M, Schieder O, Straub J, Schell J (1981) Mendelian transmission of genes introduced into plants by the Ti-plasmids of *Agrobacterium tumefaciens*. Mol Gen Genet 183:209–213
Petit A, David C, Dahl GA, Ellis JP, Guyon P, Casse-Delbart F, Tempe J (1983) Further extension of the opine concept: plasmids in *Agrobacterium rhizogenes* cooperate for opine degradation. Mol Gen Genet 190:204–214
Pfeiffer P, Hohn T (1983) Involvement of reverse transcription in the replication of the plant virus CaMV: a detailed model and test of some aspects. Cell 33:781–784
Rao SS, Lippincott BB, Lippincott JA (1982) *Agrobacterium* adherence involves the pectic portion of the host cell wall and is sensitive to the degree of pectin methylation. Physiol Plant 56:374–380
Ream LW, Gordon MP (1982) Crown gall disease and prospects for genetic manipulation of plants. Science (Wash DC) 218:854–859
Ream LW, Gordon MP, Nester EW (1983) Multiple mutations in the T-region of *Agrobacterium tumefaciens* tumor-inducing plasmid. Proc Natl Acad Sci USA 80:1660–1664
Rigby PWJ (1983) Cloning vectors derived from animal viruses. J Gen Virol 64:255–266
Risuleo G, Battistoni P, Constantino P (1982) Regions of homology between tumorigenic plasmids from *Agrobacterium rhizogenes* and *Agrobacterium tumefaciens*. Plasmid 7:45–51
Rosenberg M, Court D (1979) Regulatory sequences involved in the promotion and termination of RNA transcription. Annu Rev Genet 13:319–353
Ruvkun GB, Ausubel FM (1981) A general method for site directed mutagenesis in prokarystes. Nature (Lond) 289:85–88
Sacristan MD, Melchers G (1977) Regeneration of plants from "habituated" and "*Agrobacterium*-transformed" single cell clones of tobacco. Mol Gen Genet 152:111–117
Salomon F, Deblaere R, Leemans J, Hernalsteens J-P, Van Montagu M, Schell J (1984) Genetic identification of functions of TR-DNA transcripts in octopine crown galls. EMBO J 3:141–146
Schröder J, Schröder G, Huisman H, Schilperoort RA, Schell J (1981) The mRNA for lysopine dehydrogenase in plant tumor cells is complementary to a Ti-plasmid fragment. FEBS Lett 129:166–168
Schröder G, Schröder J (1982) Hybridization selection and translation of T-DNA encoded in RNAs from octopine tumors. Mol Gen Genet 185:52–55
Schröder G, Klipp W, Hillebrand A, Ehring R, Koncz C, Schröder J (1983) The conserved part of the T-region in Ti-plasmids expresses four proteins in bacteria. EMBO J 2:403–409
Schröder G, Waffenschmidt S, Weiler EW, Schröder J (1984) The T-region of Ti-plasmids codes for an enzyme synthesizing indole-3-acetic acid. Eur J Biochem 138:387–391
Shaw CH, Leemans J, Shaw CH, Van Montagu M, Schell J (1983) A general method for the transfer of cloned genes to plant cells. Gene (Amst) 23:315–330
Simpson RB, O'Hara PJ, Kwok W, Montoya AM, Lichtenstein C, Gordon MP, Nester EW (1982) DNA from the A6 S/2 crown gall tumor contains scrambled Ti-plasmid sequences near its junctions with plant DNA. Cell 29:1005–1014
Skoog F, Miller CO (1957) Chemical regulation of growth and organ formation in plant tissues cultured in vitro. Symp Soc Exp Biol 11:118–131
Smith GR (1983) Chi hotspots of generalized recombination. Cell 34:709–710
Sutton WD, Gerlach WL, Schwartz D, Peacock WJ (1984) Molecular analysis of Ds controlling element mutations at the Adh 1 locus of maize. Science (Wash DC) 223:1265–1268
Tate ME, Ellis JG, Kerr A, Tempe J, Murray KE, Shaw KJ (1982) Agropine: a revised structure. Carbohydr Res 104:105–120
Temin HM (1980) Origin of retroviruses from cellular movable genetic elements. Cell 21:599–600

Tempe J, Petit A, Holsters M, Van Montagu M, Schell J (1977) Thermosensitive step associated with transfer of the Ti-plasmid during conjugation: possible relation to transformation in crown gall. Proc Natl Acad Sci USA 74:2848–2849

Thomashow MF, Nutter R, Montoya AL, Gordon MP, Nester EW (1980a) Integration and organization of Ti plasmid sequences in crown gall tumors. Cell 19:729–739

Thomashow MF, Nutter R, Postle K, Chilton M-D, Blattner FR, Powell A, Gordon MP, Nester EW (1980b) Recombination between higher plant DNA and the Ti plasmid of *Agrobacterium tumefaciens*. Proc Natl Acad Sci USA 77:6448–6452

Thomashow MF, Knauf VC, Nester EW (1981) The relationship between the limited and wide host range octopine type Ti plasmids of *Agrobacterium tumefaciens*. J Bacteriol 146:484–493

Toh H, Hayashida H, Miyata T (1983) Sequence homology between retroviral reverse transcriptase and putative polymerases of hepatitis B virus and cauliflower mosaic virus. Nature (Lond) 305:827–829

Turgeon R (1982) Teratomas and secondary tumors. In: Kahl G, Schell J (eds) Molecular biology of plant tumors. Academic, New York, pp 391–414

Turgeon R, Wood HN, Braun AC (1976) Studies on the recovery of crown gall tumor cells. Proc Natl Acad Sci USA 73:3562–3564

Urisic D, Slighton JL, Kemp JD (1983) *Agrobacterium tumefaciens* T-DNA integrates into multiple sites of sunflower crown gall genome. Mol Gen Genet 190:494–508

Van Haute E, Joos H, Maes M, Warren G, Van Montagu M, Schell J (1983) Intergeneric transfer and exchange recombination of restriction fragments cloned in pBR322: a novel strategy for the reversed genetics of the Ti-plasmids of *Agrobacterium tumefaciens*. EMBO J 2:411–417

Van Larebeke N, Engler G, Holsters M, Van den Elsacker S, Zaenen I, Schilperoort RA, Schell J (1974) Large plasmid in *Agrobacterium tumefaciens* essential for crown-gall inducing ability. Nature (Lond) 252:169–170

Van Slogteren GMS, Hoge JHC, Hooykaas PJJ, Schilperoort RA (1983) Clonal analysis of heterogeneous crown gall tumor tissues induced by wild type and shooter mutant strains of *Agrobacterium tumefaciens* – expression of T-DNA genes. Plant Mol Biol 2:321–333

Velten J, Willmitzer L, Leemans J, Ellis J, Deblaere R, Van Montagu M, Schell J (1983) T_R genes involved in agropine production. In: Pühler A (ed) Molecular genetics of the bacteria-plant interaction. Springer, Berlin Heidelberg New York, pp 292–302

Walden RM, Howell SH (1982) Intergenomic recombination events among pairs of defective cauliflower mosaic virus genomes in plants. J Mol Appl Genet 1:447–456

Walden RM, Howell SH (1983) Uncut recombinant plasmids becoming nested cauliflower mosaic virus genome infect plants by intragenomic recombination. Plant Mol Biol 2:27–31

Watson B, Currier TC, Gordon MP, Chilton M-D, Nester EW (1975) Plasmid required for virulence of *Agrobacterium tumefaciens*. J Bacteriol 123:255–264

Weiler EW, Spanier K (1981) Phytohormones in the formation of crown gall tumors. Planta (Berl) 153:326–337

White FF, Nester EW (1980a) Hairy root: plasmid encodes virulence traits in *Agrobacterium tumefaciens*. J Bacteriol 144:710–720

White FF, Nester EW (1980b) Relationship of plasmids responsible for hairy root and crown gall tumorigenicity. J Bacteriol 144:710–720

White FF, Ghidossi G, Gordon MP, Nester EW (1982) Tumor induction by *Agrobacterium rhizogenes* involves the transfer of plasmid DNA to the plant genome. Proc Natl Acad Sci USA 79:3193–3197

Willmitzer L, De Beuckleer M, Lemmers M, Van Montagu M, Schell J (1980) The Ti-plasmid-derived T-DNA is present in the nucleus and absent from plastids of plant grown crown gall cells. Nature (Lond) 287:359–361

Willmitzer L, Otten L, Simons G, Schmalenbach W, Schröder J, Schröder G, Van Montagu M, De Vos G, Schell J (1981a) Nuclear and polysomal transcripts of T-DNA in octopine crown gall suspension and callus cultures. Mol Gen Genet 182:255–262

Willmitzer L, Schmalenbach W, Schell J (1981b) Transcription of T-DNA in octopine and noplaine crown gall tumours is inhibited by low concentrations of α-amanitin. Nucleic Acids Res 9:4801–4812

Willmitzer L, Sanchez-Serrano J, Buschfeld E, Schell J (1982a) DNA from *Agrobacterium rhizogenes* is transferred to and expressed in axenic hairy root plant tissues. Mol Gen Genet 186:16–22

Willmitzer L, Simons G, Schell J (1982b) The TL-DNA in octopine crown gall tumors codes for seven well defined polyadenylated transcripts. EMBO J 1:139–146

Willmitzer L, Dhaese P, Schreier PH, Schmalenbach W, Van Montagu M, Schell J (1983) Size, location and polarity of T-DNA encoded transcripts in nopaline crown gall tumors; common transcripts in octopine and nopaline tumors. Cell 32:1045–1056

Winter JA, Wright RL, Gurley WB (1984) Map locations of five transcripts homologous to T_R-DNA in tobacco and sunflower crown gall tumors. Nucleic Acids Res 12:2391–2406

Woolston CJ, Covey SN, Penswick JR, Davies JW (1983) Aphid transmission and a polypeptide specified by a defined region of the cauliflower mosaic virus genome. Gene (Amst) 23:15–23

Wöstemeyer A, Otten LABM, Schell J (1984) Sexual transmission of T-DNA in abnormal tobacco regenerants transformed by octopine and nopaline strains of *Agrobacterium tumefaciens*. Mol Gen Genet 194:500–507

Wullems GJ, Molendijk L, Ooms G, Schilperoort RA (1981a) Differential expression of crown gall tumor markers in transformants obtained after in vitro *Agrobacterium tumefaciens*-induced transformation of cell wall regenerating protoplasts derived from *Nicotiana tabacum*. Proc Natl Acad Sci USA 78:4344–4348

Wullems GJ, Molendijk L, Ooms G, Schilperoort RA (1981b) Retention of tumor markers in F1 progeny plants from in vitro induced octopine and nopaline tumor tissues. Cell 24:719–727

Wullems GJ, Krens FA, Schilperoort RA (1983) Plant protoplast transformation by *Agrobacterium tumefaciens* and its Ti-plasmid DNA. In: Pühler A (ed) Molecular genetics of the bacteria-plant interaction. Springer, Berlin Heidelberg New York, pp 274–283

Xiong C, Muller S, Lebeurier G, Hirth L (1982) Identification by immunoprecipitation of cauliflower mosaic virus in vitro major translation product with a specific serum against viroplasm protein. EMBO J 1:971–976

Yadav NS, Postle K, Saiki RK, Thomashow MF, Chilton M-D (1980) T-DNA of a crown gall teratoma is covalently joined to host plant DNA. Nature (Lond) 287:458–461

Yadav NS, Vanderleyden J, Bennett DR, Barnes WM, Chilton M-D (1982) Short direct repeats flank the T-DNA on a nopaline Ti-plasmid. Proc Natl Acad Sci USA 79:6322–6326

Yamaoka N, Furusawa II, Yamamoto M (1982) Infection of turnip protoplast with cauliflower mosaic virus DNA. Virology 122:503–505

Yang F, Simpson RB (1981) Revertant seedlings from crown gall tumors retain a portion of the bacterial Ti-plasmid DNA sequences. Proc Natl Acad Sci USA 78:4151–4155

Yang F, Montoya AL, Merlo DJ, Drummond MH, Chilton MD, Nester EW, Gordon MP (1980) Foreign DNA sequences in crown gall teratomas and their fate during the loss of the tumorous traits. Mol Gen Genet 177:707–714

Zambryski P, Holsters M, Krüger K, Depicker A, Schell J, Van Montagu M, Goodman HM (1980) Tumor DNA structure in plant cells transformed by *A. tumefaciens*. Science (Wash DC) 209:1385–1391

Zambryski P, Depicker A, Krüger K, Goodman HM (1982) Tumor induction by *Agrobacterium tumefaciens:* analysis of the boundaries of T-DNA. J Mol Appl Genet 1:361–370

Zambryski P, Joos H, Genetello C, Leemans J, Van Montagu M, Schell J (1983) Ti-plasmid vector for the introduction of DNA into plant cells without alteration of their normal regeneration capacity. EMBO J 2:2143–2150

Zambryski P, Herrera-Estrella L, De Block M, Van Montagu M, Schell J (1984, in press) The use of the Ti-plasmid of *Agrobacterium* to study the transfer and expression of foreign DNA in plant cells: new vectors and methods. In: Setlow J, Hollaender A (eds) Genetic engineering, principles and methods, vol 6. Plenum, New York

Epilogue

By H. BINDING and J. REINERT

In collaboration with the first authors of this volume

Isolation, as well as fusion of somatic cell protoplasts and transplantation of genes, are all manipulations which alter the metabolic and developmental patterns of plant cells. Investigations with cell lines derived from such manipulated cells have already revealed information on the physiological control of the developmental steps, on the developmental potencies of protoplasts with respect to their source and genotype, on cytoplasmic incompatibility and related phenomena, and on the integration and developmental expression of foreign genes.

Various types of higher plant cell are capable of starting new developmental pathways when they are isolated from their native environment in the organized tissue and the new developmental processes are susceptible to exogeneous factors. Interestingly, there are indications that plant cells may also be irreversibly switched to a particular differentiation pathway in their early development. These characters, the plasticity and supposed determination, make isolated protoplasts and fusion products of higher plants well suited for the investigation of the following problems:

(1) Which types of cell can be diverted from their in situ differentiation pattern? (2) Does the lack of plasticity depend on genetic or epigenetic events? (3) What new developmental pathways can be induced in various types of cells? (4) How stable are the induced developmental pathways? (5) What factors control direction and single steps of development? (6) What correlations exist between the taxonomic position of a plant and the responses of its cells to in vitro culture conditions? (7) Which mechanisms of physiological incompatibility are realized in plants and which are utilized in sex, parasitism, and root anastomosis?

Recent advances in plant molecular biology are contributing and will continue to contribute greatly to developmental studies. Genetic engineering of foreign, dominant selectable genes into plant cells now provide useful markers for cellular manipulations such as cell fusion. These transformation vectors can also be used to introduce other nonselectable genes into plants. This, in conjunction with our ability to isolate specific plant genes and manipulate them in vitro, provides us with a powerful tool to study the molecular basis of developmental expression of the introduced genes. We can hope to see rapid advances in the identification of the regulatory sequences of different plant genes that are responsible for their developmental expression.

While the molecular, genetic, and cellular approaches to development outlined above will contribute greatly to our basic knowledge of control of development and differentiation, it will be equally important for practical applications.

Results and Problems in Cell Differentiation 12
Differentiation of Protoplasts and of Transformed Plant Cells (Edited by J. Reinert and H. Binding)
© Springer-Verlag Berlin Heidelberg 1986

Protoplast regeneration is an excellent tool for single cell cloning in order to propagate certain genotypes and to dissociate somaclonal variants, to screen the response to various factors of the environment and to viruses, to obtain virus-free plants and new types of mutant. Plants with new heritable combinations may be constructed by protoplast fusion and gene transplantation. The range of application of somatic cell fusion includes mainly the addition of unreduced genomes within or between close-related species in order to obtain controlled heterosis; the exchange of selected genetic traits between sexually incompatible species after unilateral loss of chromosomes and consecutive backcrossing; and the substitution and recombination of cell organelles. The combination of genes ranging from remote species to members of different kingdoms is made feasible by gene technological approaches.

Only a few crop species are presently amenable to protoplast techniques and genetic manipulations through cell and tissue culture. If we are to genetically improve these crop species by the new technologies, we will have to develop methods to overcome their inability to regenerate from protoplasts or tissues in culture. Even for plants that can be regenerated, we need to study tissue-specific expression of foreign genes, especially from taxonomically remote species, the effect of the site of integration in the chromosome on the expression of the foreign gene, and the stability of the foreign gene. Progress in the respective investigations has been rapid. This holds true, for instance, with *Agrobacterium*-mediated gene transfer (see Part IV), and transformation by free DNA (see Lörz et al. 1985; Potrykus et al. 1985a, b).

References

Lörz H, Baker B, Shell J (1985) Gene transfer to cereal cells mediated by protoplast transformation. Molec Gen Genet 199:178–182

Potrykus I, Pszkowski J, Saul MW, Petruska J, Shillito RD (1985a) Molecular and general genetics of a hybrid foreign gene introduced into tobacco by direct gene transfer. Mol Gen Genet 199:169–177

Potrykus I, Saul MW, Petruska J, Paszkowski J, Shillito RD (1985b) Direct gene transfer to cells of a graminaceous monocot. Mol Gen Genet 199:183–188

Appendix

1 Abbreviations

A2CA[r]	algo-*a*zetidine-*2*-*c*arboxylic *a*cid *r*esistant mutant
AEC[r]	S-(2-*a*minoethyl)-L-*c*ystein *r*esistance mutant
APH	*a*minoglycoside *ph*osphotransferase
ars	*a*utonomously *r*eplicating *s*equence
B5	a plant tissue culture medium (Gamborg et al. 1968)
bp	*b*ase *p*airs of DNA
CaMV	*ca*uliflower *m*osaic *v*irus
cms	*c*ytoplasmic *m*ale *s*terility, controlled by mitochondria
2,4-D	*2,4-d*ichlorophenoxyacetic acid
DPD	a protoplast culture medium (*D*urand et al. 1973)
F5	a protoplast culture medium (*F*rearson et al. 1973)
G-418	a 2-deoxystreptamine antibiotic which inhibits protein synthesis
kd	*k*ilo*d*alton
KM	the protoplast culture medium 8p (*K*ao and *M*ichayluk 1975)
mtDNA	*mit*ochondrial *DNA*
NAA	α-*n*aphthalene *a*cetic *a*cid
NR^-	*n*itrate *r*eductase deficient mutant
ORF	*o*pen *r*eading *f*rame
PEG	*p*oly*e*thylene *g*lycol
ptDNA	*plast*id *DNA*
R67	a *r*esistance plasmid
R77	a *r*esistance plasmid
R prime	derivatives of *R* plasmids, for example a cointegrate of R and Ti plasmid
Ri	*r*oot *i*nducing plasmid of *Agrobacterium rhizogenes*
Roi	*r*oot *i*nhibiting – locus on T-DNA
RUBPC	*r*ib*u*lose-1,6-*b*is*p*hosphate *c*arboxylase
Shi	*sh*oot *i*nhibiting – locus on T-DNA
Su	semidominant *su*lphur mutant in tobacco
SV 40	*s*imian *v*irus 40
T-DNA	*t*ransforming sequence of the Ti plasmid
Ti	*t*umor *i*nducing plasmid of *Agrobacterium tumefaciens*
T_L-DNA	obligate sequence of *T*-DNA when an adjacent sequence of the Ti plasmid (the T_R region) is additionally found in an octopine tumor

Tms	*t*umor *m*orphology *s*hoot locus (same as Shi) on T-DNA
Tmr	*t*umor *m*orphology *r*ooty locus (same as Roi) on T-DNA
Tml	*t*umor *m*orphology *l*arge locus on T-DNA of octopine tumors
Tn	transposable element, *t*ranspose*n*
T_0	a protoplast culture medium (Carboche 1980)
V-47	a protoplast culture medium (Binding 1974)
V-KM	a protoplast culture medium combining components of *V*-47 and *KM* (Binding and Nehls 1977)
vir	*v*irulence region of the Ti plasmid

2 Glossary

A number of terms are used in this book which are not equally familiar to all readers of the series. In most cases it is sufficient to consult text books in botany or genetics. In genetics, the *Glossary of Genetics and Cytogenetics* (Rieger et al. 1976) is, furthermore, recommended. The terminology in tissue culture has been recently published by a Terminology Committee of the Tissue Culture Association (Schaeffer et al. 1984). Criteria for the selection of terms in the following list were their specialized and particular interest with respect to the topics treated in this volume, as well as their utilization in a sense which deviates from the common definition or from a special definition given by Schaefer et al.

adventitious organ formation – organization of shoots or parts of shoots from callus or organs other than shoot apical and axillary meristems, of roots from callus or organs unlike roots, and of embryos from somatic cells (sometimes named embryoids)

agglutination – adherence of isolated plant protoplasts to one another after removal of the negative surface charges of their plasmalemmata

budding – local, bubble-shaped swelling of isolated protoplasts during cell wall regeneration

cell cluster – colony of less than about 50 adjacent, congenital cells

cell density – number of cells in suspension per 1 ml

chimeric gene – a gene which is constructed of functional units of different origin, which combines, for instance, the promoter of one gene with the coding sequence of another gene

co-culture – culture of mixtures of two types of protoplast, one of them supporting the growth of the other type by conditioning the media

conditioning – improvement of the properties of culture media by the action of cells. The conditioning cells may be removed before utilization of the media. Conditioning during the culture of the cells to be promoted is obtained by appropriate cell densities, feeder layers, or co-culture

crown gall – a plant tumor induced by infection and transformation by *Agrobacterium tumefaciens*

cybrid – a cell/organism in which the nucleus of one species is associated with plastids and/or mitochondria containing heterospecific or recombinant genophores. The term is used irrespective of whether *cybridization* was obtained by fusing functionally incomplete protoplasts, or if it arose from nucleus segregation during development of a complete fusant

cytoplast – a subprotoplast lacking a nucleus

differentiation – the formation of a cell which is organized unlike a meristem cell. The term is also used with respect to organ formation of callus

feeder layer – solidified nutrient medium containing X-ray-inactivated cells/protoplasts which is used in order to condition a culture put into an upper layer

fusion of somatic cell protoplasts – the establishment of cytoplasmic community between protoplasts. The *fusion product (fusant)* is assigned *fusion body* as long as no rigid cell wall is regenerated. Fusion is signed by (x)

hairy root disease – abnormal root production of dicotyledonous plants caused by infection and subsequent transformation by *Agrobacterium rhizogenes*

heterokaryocyte – a heterokaryotic plastocyte developed from a fusion body
mesophyll – leaf parenchyma cells (predominantly palisade and sponge parenchyma). Mesophyll protoplast preparations, however, may also contain protoplasts of epidermal and bundle cells
miniprotoplast – a subprotoplast containing a nucleus
opines – novel metabolites which are formed specifically in plant cells after infection by *Agrobacterium*. The responsible genetic information is located on the T-DNA. The following opines and derivatives are mentioned in part IV: *nopaline, octopine, agropine, agrocimopine,* and *succinamopine*
plantlet – a small plant under tissue culture conditions
plastocyte – the cell which developed from an isolated protoplast by cell wall regeneration
plating density – cell density at the time of plating
plating efficiency – percentage of plated cells which formed colonies. In protoplast culture, it is often used to indicate the yield of protoplasts which were able to form at least bicellular clusters
protoclone – the population of cells/individuals which developed vegetatively from a single protoplast or fusion body
protoplast – the body of the cell within the cell wall. It is detached from the wall by slight dehydration (plasmolysis). The term is also used for the *isolated protoplast*
protoplast regeneration – the development of plastocytes (cell wall regeneration) and plantlets from isolated protoplasts. The term is used by some authors exclusively with respect to plant regeneration from protoplasts
shoot culture – a type of tissue culture in which shoots are propagated by subculturing shoot tips and – if possible – promoting adventitious shoot formation. Callus and root formation is suppressed as far as possible
somaclonal variation – the formation of mosaic tissue by the genesis of variant cell lines by mutational events
spheroplast – a prokaryotic cell after removal of the rigid cell wall components
spontaneous fusion – fusion of neighbor cells during digestion of cell walls
subprotoplast – an osmotically entire protoplast fragment (see cytoplast and miniprotoplast)
teratoma – a plant tumor producing rudimentary shoots
transformation – introduction of foreign genetic material into cells
transposon mutagenesis – inactivation of a gene by the insertion of a transposon

References

Rieger R, Michaelis A, Green MM (1976) A glossary of genetics and cytogenetics. Springer, Berlin
Schaeffer WI, Makino RK, Moehring TJ, Sanford K, Schneider I (1984) Usage of vertebrate, invertebrate and plant cell, tissue and organ culture terminology. In vitro 20:19–24

Index of Scientific Names of Bactaria, Fungi, and Plant Genera

English names and other taxa are not listed.
Hyphen means that the respective genus is treated sporadically on any of the comprised pages.

Subject Index